D0794563

# Chemistry for the Million

# Chemistry for the Million

Richard Furnald Smith

Charles Scribner's Sons · *New York*

A—7.72(V)

PRINTED IN THE UNITED STATES OF AMERICA

Library of Congress Catalog Card Number 77-37219
SBN 684-12771-7 (trade cloth)

It all comes down to chemistry in the end, of course. Hydrogen, oxygen. . . . What are the other things? God, how infuriating, how infuriating not to know!

——Aldous Huxley, *Point Counter Point*

# Contents

# Preface

Anyone who can do a crossword puzzle can learn many of the basic ideas of chemistry. The results may also be more rewarding—instead of filling his head with mutually irrelevant scraps of lore, he will be following a sustained effort of the human race to make sense of the material world.

"A man who writes a preface rarely has a tranquil conscience," said Jean Anouilh; "he has something to confess." What the writer on science has to confess is clear enough. He has exercised choice; he has simplified. In a preface he can acknowledge his omissions and compromises. This book is not a textbook, or a history of the science. It is intended for the reader—puzzle fan or not—who would like to know something about chemistry and its past. If it succeeds in arousing an interest in the subject, the Suggestions for Further Reading will provide guidance in pursuing it.

Not all the secrets of chemistry are contained in this volume. After each chapter the reader should consider the following three axioms, which also apply to books written by specialists for specialists:

1. It is not that simple.
2. It is not up to date.
3. Someone somewhere disagrees with it.

# Chemistry for the Million

# 1. Ancient Origins

Chemistry has been a science in the modern sense for only two or three centuries. For twenty centuries before that it was inseparable from sorcery and astrology. It began as simple technology: metallurgy, textile dyeing, with a little beer making on the side.

Fire was man's first source of chemical energy. Ashes and charred animal bones were found with the *Homo erectus* fossils near Peking. Apparently man learned to keep a fire going about two million years ago, in the Middle Pleistocene Age—long before he learned to start one. This is a skill no other primate has acquired, or shown any interest in. By maintaining a fire in the mouth of a cave, early man could keep wild animals away, warm the cave, and leave the women and children protected inside when he went out hunting.

The management of fire was the beginning of chemistry.

When man saw that certain soils grew hard near the fire, he began to make pottery and bricks. When he noticed that shiny materials sometimes melted out of heated rocks, he began to work with metals.

Using fire, man was able to extract gold, silver, and copper. These softer metals have low chemical activity, so they are usually found in a pure or easily purifiable state. (Active metals such as aluminum were not purified until quite recently.) By 1200 B.C. man was also smelting lead, tin, iron, and mercury.

A major breakthrough occurred with the discovery that adding a little tin to melted copper produced something much tougher than copper or tin alone: bronze. Whole empires were built around this technological superiority—from the Tigris-Euphrates valley and Asia Minor to Egypt, mainland Greece, and Western Europe. A prehistoric mine near Salzburg is estimated to have produced 20,000 metric tons of copper during the Late Bronze Age.

But there is danger in what Arnold Toynbee in *A Study of History* calls the "idolization of an ephemeral technique." Upstart nations which became skilled at forging iron made short work of the older empires. Invading Dorians conquered the bronze-age Mycenaean civilization in Greece (900 B.C.); Persia took over Babylonia and Egypt (500 B.C.). Some of this is reflected in the Greek poet Hesiod's *Works and Days* (eighth century B.C.), which describes the splendid gold and silver ages being replaced by the age of bronze and finally by a race of iron users.

In addition to exploring the military and artistic potentialities of metals, early men also investigated plants, not just for their food value, but for their pharmacological properties as well. Today's most commonly prescribed asthma remedy, ephedrine, has been used for centuries by the Chinese in the form of horsetail plants (*Ephedra*). The ancient Greeks were thoroughly familiar with the physiological effects of hemlock (*Conium maculatum*). Other commonly used plant drugs included camphor, digitalis, strychnine, opium, coca, menthol, quinine, ipecac, castor oil, and belladonna.

Most cultures, from the simplest to the most complex, have used at least one plant material simply because it makes the user feel good. The Polynesians had kava kava, North American Indians had peyote: we have whisky.

Eventually the study of plant properties developed into medicine, while metallurgy turned into alchemy. Modern chemistry is a fusion of these two ancient empirical sciences.

# 2. Alchemy

*It is fun to follow the growth of science out of magic and demonology, and see it declining again in our time back into magic, its parent.*
——Lawrence Durrell, *Spirit of Place*

Alchemical ideas seem to have arisen independently in China, India, the Near East, and the West. The early stages of Western alchemy are closely tied to the city of Alexandria.

More than three centuries before Christ, Aristotle (384–322 B.C.) described the cosmos as being made of four imperfect elements: air, fire, earth, and water. A fifth, perfect element—ether, the *quinta essentia,* or "quintessence"—permeated the other four, like a kind of soul of the universe. These ideas have come down more or less intact in astrology as the diagram on the next page shows.

The Greek physicians Hippocrates (c.460–377 B.C.) and Galen (A.D. 129–199) extended this scheme to human physiology. The body contained four basic juices, or humors, corresponding to the four elements:

blood—air
yellow bile (from the liver)—fire
black bile (from the spleen)—earth
phlegm (from the brain)—water

One remained healthy by keeping these humors in balance. According to which humor was strongest, one's temperament was sanguine, choleric, melancholic, or phlegmatic. Melancholy is actually Greek for "black bile." Its association with the spleen has persisted over the centuries; "Spleen" was the title given by the nineteenth-century French poet Charles Baudelaire to three of the gloomiest poems in *Les Fleurs du Mal.*

The great German philosopher Immanuel Kant (1724–1804) wrote shrewd psychological descriptions of the four temperaments. Some psychologists continue to find the system useful: H. J. Eysenck relates it to introversion/extraversion and stability/instability and in *Fact and Fiction in Psychology*

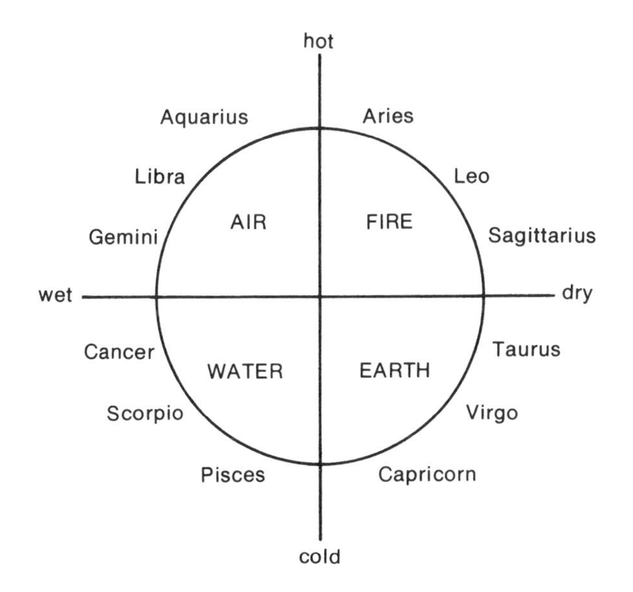

(1965) supplies eight experimentally determined adjectives for each type. Here are the original four humors and temperaments with some of Eysenck's additions:

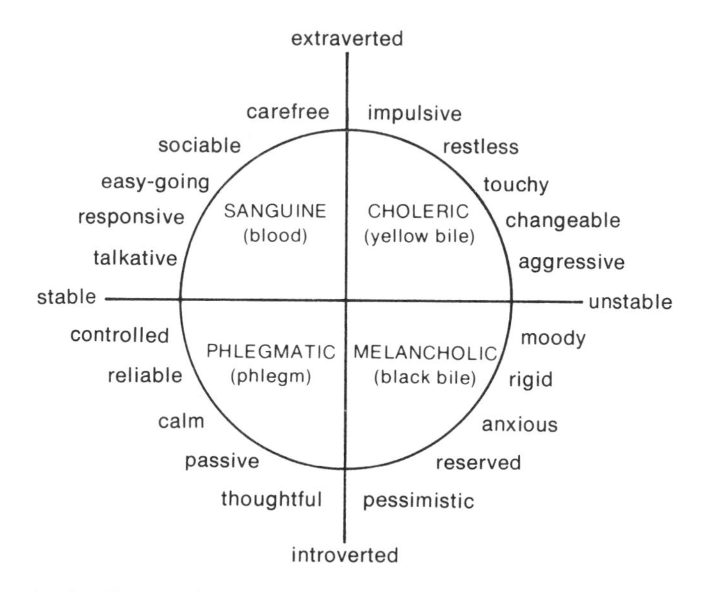


Artists have often used the system of humors. Paintings and engravings of the four temperaments were common in the Renaissance (Albrecht Dürer's *Melencolia,* for example). Each movement of Carl Nielsen's Symphony No. 2 represents a different temperament (inspired by paintings seen in a Danish bar). So does each variation in Paul Hindemith's Theme and Variations for Piano and Strings, which was used by George Balanchine for one of his most austerely beautiful ballets, *The Four Temperaments*. (Incidentally, Nielsen starts out choleric and ends sanguine, while Hindemith starts melancholic and ends choleric.)

Aristotle did not invent the four-square, earth-air-water-fire theory of matter. He merely organized the prevalent beliefs of his day. These included the idea that one element could be converted into another under certain conditions. For example, air could be converted to water by overcoming its hotness with coldness.

Aristotle's prize pupil, of course, was Alexander the Great. After conquering Egypt (332 B.C.), Alexander founded a city on the westernmost mouth of the Nile, to be a link between East and West. What could have been more natural than for him to name the city Alexandria, after himself?

The new rulers of Egypt and heirs to Alexander—the Ptolemies—were Macedonian Greeks. While adopting certain Egyptian customs (such as brother-and-sister marriage), they did not bother to learn the Egyptian language and seldom ventured beyond Alexandria. They did not intermarry with Egyptians. Two centuries after the Roman conquest (31 B.C.), Alexandrian scientists still wrote in Greek. Even an astrological treatise like the famous *Tetrabiblos* by the astronomer Claudius Ptolemy (A.D. 85–165) is very much in the Greek rationalist tradition, with no hint of sensational Egyptian mysteries.

Thousands of Greeks migrated to Alexandria, attracted by Ptolemies' offer of free land. Many of the ancient world's most celebrated scholars settled there. Alexandria became a center of intellectual ferment, famous for its street riots and for its library, the largest in the ancient world. Hellenistic, Jewish, and Egyptian cultures flourished side by side.

A favorite recreation of the Greek philosophers there was watching the city's artisans at work. Perfumes, cosmetics, glass, textiles, and jewelry were manufactured in Alexandria and exported throughout the Mediterranean world. Some artisans were especially skillful at imitating precious stones and making cheap metals look expensive. The philosophers saw commonplace metals going into the foundries and an apparent treasure trove of gold and silver emerging. Needless to say, the artisans were not anxious to reveal their trade secrets. If a group of simple-minded academics chose to believe that base metals were being transformed into gold, let them.

So began a curious association between practical artisans and neo-Platonists, with characteristic overtones of Egyptian magic and charlatanism. It was still flourishing in A.D. 642 when the Arabs conquered Alexandria. The word alchemy actually comes from the Arabic "Al Kimia." The "al" is like the Spanish "el" (the), which it later gave rise to. The "kimia" may be from a seldom used term for Egypt (making alchemy "the Egyptian art"), but more likely it comes from the Greek "chyma," molten metal.

Papyri from around A.D. 300 still exist. Some contain practical chemical recipes for imitating silver and gold; others claim genuine transmutations. One of the oldest alchemical documents is the *Emerald Table of Hermes Trismegistus*. It has been traced (by scholarly inference) as far back as the fourth century A.D. and seems to be Syrian rather than Alexandrian in origin. The Arabs attributed it to Apollonius of Tyana, a wandering magician of the first century.

Short and cryptic, the *Emerald Table* hints that the material world can be controlled and transformed. Approval of astrology is deduced from the line: "What is below is like what is above, and what is above is like that which is below . . ."

The "thrice-great" Hermes referred to in the title is the supposed author. The original Hermes, the Greek god of science and communication (and thieves), was equated with the Egyptian god Thoth. From this pseudonym, which was also used on other books, alchemy became known as "the Hermetic art." Other *noms de plume* popular with alchemists included Moses, Cleopatra, and Isis.

The Arabs brought with them to Spain an alchemical lore far more sophisticated than anything then known in Europe. Medical books written by al-Rāzī of Teheran (844–926) were still being studied at European universities in the seventeenth century. Al-Rāzī's *Book of the Secret of Secrets* (*Kitāb sirr al-asrār*) is actually a detailed laboratory manual, complete with lists of required equipment and chemicals, along with instructions for carrying out filtration, crystallization, and distillation. He was the first to use the classification, "animal, vegetable, or mineral." He never doubted that all substances were composed of Aristotle's four elements, and he seems to have taken it for granted that one could turn base metals into gold and "improve" glass or quartz into rubies, emeralds, and sapphires by means of various "elixirs."

The Arab physician Ibn Sīnā (980–1037), known in Europe as Avicenna, a Latinized form of the Hebrew Aven Sina, wrote knowledgeably about musical theory, physics, astronomy, and the origin of nervous disorders. His *Canon of Medicine* (*Qânûn fi'l tibb*) elaborates on the four-humors theory, lists some 760 pharmacological agents, and warns against the spread of disease by soil, water, and air. Translated into Latin as *Canon medicinae Avicennae,* this work was used in Europe for the next six centuries.

Concerning alchemy, Avicenna was a complete skeptic. He did not believe that metals had been transmuted in the past or could be in the future. He would acknowledge only the possibility that "such a degree of accuracy in imitation may be reached as to deceive even the shrewdest."

But the alchemists continued their search with undiminished enthusiasm. Distillation, which may have been known to Aristotle and to the Babylonians, was practiced by the alchemists of Alexandria and introduced into Europe by the Arabs as early as the twelfth century. By this means a number of substances were obtained in concentrated form for the first time: alcohol, nitric and sulfuric acids, a variety of essential oils from plants. The richness of their discoveries began to tax the simple four-square scheme of Aristotle. For example, there were too many different kinds of "water": *aqua fortis* (nitric acid), *aqua ardens* (alcohol), *aqua calcis* (limewater), *aqua regia* (nitric and hydrochloric acids), and so forth.

Distilled spirits were a godsend to the merchants and wine shippers. Storage problems were eliminated. Taxes, duties, and transport charges were reduced. With the discovery of the New World, "hard liquor" became even more important. The condition of a fine wine after being shaken up for several months at sea was appalling. Moreover, it was absurd to fill a small ship with a cargo that was 90 percent water. But brandy, rum, and gin were quite another story. They were concentrated, and they tasted the same at the end of a long ocean voyage as they did at the start.

The increased production of alcohol soon led to an increasing production of alcoholics. One of the most brilliant of these was the Swiss physician Theophrastus Bombast von Hohenheim (1493–1541), who renamed himself "Paracelsus," or "better than Celsus." (Celsus was a Roman physician of the first century A.D. and the author of *De Re Medica.*) Paracelsus's addictive drinking seems to have played a role in his compulsive boastfulness, his vagabond way of life, his odd literary style, and his sudden death. (When his remains were exhumed during the nineteenth century, his skull was found to be fractured.)

Perhaps inspired by Martin Luther's burning of the Papal Bull in 1520, Paracelsus began a teaching engagement in Basel seven years later by publicly burning the works of Avicenna and Galen. He attacked members of the medical establishment with great gusto. Although he often ridiculed astrology, he apparently never bled a patient or gave an enema unless the moon was in a favorable—that is, "watery"—sign of the zodiac.

He dictated some 300 books in a kind of frenzy, never bothering to reread or correct, hardly pausing to sleep. They are written in a coarse Swiss-German dialect, interspersed with schoolboy Latin, Hebrew, Greek, and Arabic, plus words of his own invention. Only a few were published during his lifetime.

After his death, his collected works went through many editions and were extremely influential. In 1599 they were all placed on the Roman Catholic Church's Index of Prohibited Books. They are a startling mixture of keen insights and superstitious twaddle. While using alchemical methods in the preparation of medicines, Paracelsus considered therapy more important than the transmutation of metals. He regarded the human body as a chemical system to be cured by mineral rather than herbal remedies. And he impressed subsequent generations with the exciting possibility that Aristotle and Galen might not have all the answers.

Many of the alchemists were sincere mystics who saw the changing of lead into gold as an allegory of the soul's transformation through enlightenment. Other alchemists were prepared to accept the ability to make gold in place of the enlightenment. Still others were willing to fake both processes. Conditions must have been ideal for confidence men: universal scientific ignorance, combined with a religion that demanded a belief in the miraculous. Increasingly well-documented cases of fraud discredited alchemy, while discoveries of the realities of chemical change made it unnecessary.

As late as 1897 and 1898, the United States government purchased bricks of "alchemical gold" at almost monthly intervals. They were the work of an English-born alchemist, Dr. Stephen H. Emmens, a member of the American Chemical Society. He disappeared just before he was to give a widely publicized demonstration of his alchemical skills at the Paris Exposition of 1900. An assay of one of the gold bricks by the United States Mint noted that it contained the kind of impurities "constantly present in old jewelry."

Whatever its limitations as science, alchemy remains a sublime metaphor. It has given gold a symbolic value far beyond its actual economic worth, as shown by works ranging from Richard Wagner's opera *Der Ring des Nibelungen* to B. Traven's novel and John Huston's film *The Treasure of Sierra Madre*. It helped form the Faust legend, which served Christopher Marlowe, Goethe, and Thomas Mann in turn (to say nothing of the composers Boïto, Berlioz, Gounod, and Busoni). Alchemical symbols enrich the work of William Butler Yeats, Carl Jung, Jorge Luis Borges, and the painter Pavel Tchelitchew.

Today, alchemy is far from dead. In fact, it reaches a kind of apotheosis in books like *Le Matin des Magiciens* by Louis Pauwels and Jacques Bergier, 1960 (translated as *The Morning of the Magicians* and as *The Dawn of Magic*). This is a heady mixture of alchemy, ESP, diabolism, interplanetary visitors, Atlantis legends, and higher states of consciousness. One's imagination is not the poorer for having read it.

Early in the eighteenth century, the Dutch chemist Hermann Boerhaave (1668–1738) was a fierce partisan of the "new chemistry" that was struggling to free itself from alchemy. He spent years trying to turn mercury into gold by every conceivable means and then reported flatly: "The mercury remained mercury."

Yet in his 1732 textbook, *A New Method of Chemistry,* he refused to condemn "such skillful artists" as the alchemists. Too much incredulity was just as bad as too much credulity, he said, adding, "The business of a wise man is to try all things, hold fast to what is proven, never limit the power of God, nor assign bounds to nature."

# 3. The Metals

The ancient Greeks and Romans knew seven metals. They also knew seven "planets" (the five nearer planets, plus the sun and the moon). Perhaps in an effort to simplify or unify the cosmos, the ancients related each planet to a specific metal.

This is why alchemical symbols for metals and astrological symbols for planets are identical:

| *English name* | *Chemical symbol* | *Latin name* | *Alchemical symbol* | |
|---|---|---|---|---|
| gold | Au | aurum | ☉ | (Sun) |
| silver | Ag | argentum | ☽ | (Moon) |
| copper | Cu | cuprum | ♀ | (Venus) |
| iron | Fe | ferrum | ♂ | (Mars) |
| mercury | Hg | hydrargyrum | ☿ | (Mercury) |
| tin | Sn | stannum | ♃ | (Jupiter) |
| lead | Pb | plumbum | ♄ | (Saturn) |

The ancients thought there was an eighth metal, "electrum," which they associated with Jupiter. But eventually they realized that this was an alloy (mixture) of gold and silver, so Jupiter had to be content with tin.

One can see why pre-scientific people would associate gold with the sun, silver with the moon, iron with warlike Mars, and lead with slow-moving Saturn, but what has copper to do with Venus? *Cuprum* is a later form of *cyprium*. The island of Cyprus has been a source of copper for the Mediterranean since the Bronze Age; it was also a center for the worship of love goddesses (first Ishtar, then Aphrodite, and finally Venus). The symbol ♀ is still used to indicate "woman," and the word "cyprian" still vibrates with sexual overtones. But whether copper was named after the island, or the island was named after the copper found there remains beyond the reach of scholarship.

The chemical symbols for the seven metals all come from the Latin names (Fe for *Ferrum,* and so forth). Metals discovered more recently have chemical symbols that correspond directly with their modern names (Al for aluminum, Mg for magnesium, and so forth). *Hydrargyrum* is Latinized Greek for "liquid silver," or quicksilver (*hydor,* water, plus *argyros,* silver). *Plumbum* reveals the long-time association between plumbers and lead pipe. Anyone who knows a Romance language will find other connections: Argentina, *ferrovía* (Spanish for "railway"), cupremia, aureomycin, and many more.

These metals are chemical elements. They cannot be broken down into anything simpler. However, they can combine with other elements. For example:

$$4Fe + 3O_2 \rightarrow 2Fe_2O_3$$

| | | |
|---|---|---|
| Fe | = | iron, an element (metallic solid) |
| O | = | oxygen, an element (gas) |
| $Fe_2O_3$ | = | ferric oxide, a compound (reddish powder) |

Ferric oxide is ordinary rust. As a pigment, it is known as

Venetian red. As a mineral, it is called hematite. Where it is abundant, it makes soils reddish (especially in the tropics). Since it is a compound, it can be broken down again into iron and oxygen; the preceding equation can be reversed. In fact, that is the way iron is obtained from the iron ore hematite. The ore is heated with charcoal, coal, or coke (all forms of the element carbon, C), and the oxygen is carried off as carbon dioxide ($CO_2$). Unfortunately, iron ores are never pure, so that a number of other things besides carbon dioxide are emitted from blast furnaces.

More than 100 elements are known today, and roughly three-quarters of them are metals. The word metal comes from the Greek word for mine, and since not everything dug out of a mine is a metal, chemists have taken pains to define metals more precisely:

Metals have "metallic luster."
Metals are "malleable" (can be hammered into sheets).
Metals are "ductile" (can be drawn into wires).
Metals have high densities (are heavy).
Metals are good conductors of heat and electricity.

The Arab alchemists included another property in the list: Metals are "sonorous": they make good bells, cymbals, trumpets, and glockenspiels.

Gold, silver, and copper are often called the "coinage metals," because of their long history as mediums of exchange. Brief descriptions of these and other important metals follow.

*Gold (Au).* Gold has been valued since prehistoric times, not only for its beauty and rarity, but for its simple *unreactivity*. Objects made from it last for centuries without rusting or corroding. Almost all the gold in nature occurs in the free (elemental) state. The gold found in the 1898 Alaskan Gold Rush, for example, was almost completely pure. Because of its resistance to corrosion, technology continues to find new uses for gold.

The proportion of gold in an alloy is measured in karats.

Pure gold is 24 karats. Thus a 12-karat ring will contain 50 percent gold, with perhaps 35 percent copper and 15 percent silver. Gold coins are usually 90 percent gold and 10 percent copper.

*Silver* (*Ag*). Slightly more reactive than gold, this metal is especially sensitive to sulfur and will readily "tarnish" if exposed to it —that is, a black deposit of silver sulfide ($Ag_2S$) will form on its surface.

Certain compounds of silver are sensitive to light. Silver bromide (AgBr) in a layer of gelatin is the basic ingredient of photographic films.

Sterling silver is 92.5 percent silver and 7.5 percent copper. Silver coins in the United States are 90 percent silver and 10 percent copper. "German silver" contains no silver at all but is an alloy of 55 percent copper, 25 percent nickel, and 20 percent zinc.

*Copper* (*Cu*). Copper is sufficiently reactive to be found in combined forms (as $Cu_2O$ and $Cu_2S$) as well as in the free form. Copper ores tend to be distinctively colored and thus must have been easy for early man to locate.

Cuprous oxide ($Cu_2O$) has a characteristic reddish-brown color. The strikingly beautiful *sang-de-boeuf* (ox-blood) glazes on Chinese porcelain of the Sung and Ming dynasties are derived from this compound. The same compound forms in the common clinical test for diabetes when there is glucose in the urine.

Copper sulfate ($CuSO_4$) is poisonous but more poisonous to plants than to people. It is used to kill algae in swimming pools and reservoirs and to keep tree roots from invading drainpipes.

Pure copper is quite soft and can be easily worked into ornaments, but it makes a terrible kitchen knife. Addition of a little tin (Sn) produces the alloy bronze, harder than either copper or tin alone, suitable for swords, spears, armor—and, because of its low melting point, ideal for casting statues. It is less successful as jewelry; a bronze ring will turn the wearer's finger green. The composition of bronze can be varied according to the properties desired. Ordinary bronze is 90 percent copper and 10 percent tin. Bronze for church bells is 70 percent copper and 30 percent tin. Brass, an alloy of copper and zinc, is traditionally used for ship fittings and cartridge shells (70 percent copper and 30 percent zinc).

*Platinum (Pt).* Like gold, platinum is rare and extremely unreactive. Its color is silvery white, however, rather than yellow-green. Platinum laboratory ware is useful in handling liquids and gases so hyperactive that they react with nearly everything—including their own containers.

While unreactive itself, platinum is an excellent catalyst; its presence makes certain reactions "go" faster than usual, even though no platinum is permanently changed or used up by the process.

"White gold" can be made by adding a small amount of platinum to gold. It can also be made (much more cheaply) by using nickel, or nickel, zinc, and copper, instead of platinum.

*Iron (Fe).* This ancient metal is found all over the earth's surface. Except for meteorites, it is nearly always in the combined form. The colors of rocks are due almost entirely to iron compounds. Iron is the most widely used of all the metals.

Since iron is quite active chemically, however, it is subject to rust and corrosion. It reacts not only with oxygen, but with sulfur, phosphorus, chlorine, and dilute acids. A great deal of ingenuity has gone into making iron more resistant to chemical attack.

Iron can be protected simply by covering it with grease or painting it (automobiles). Ceramic enamels are used for sinks, bathtubs, and refrigerators. Iron can also be plated with another, more resistant metal. "Galvanized iron," for example, is iron coated with zinc. Nickel, aluminum, chromium, and lead are also used as protective surfacing materials. Tin cans are actually iron cans plated with tin.

In the "Oz" books by L. Frank Baum some metallurgical confusion occurs. Although made entirely of tin, the Tin Woodman, in *The Wizard of Oz,* "rusts" whenever he gets wet and must be oiled to become mobile again. In *The Land of Oz,* the Tin Woodman has himself nickel-plated: one protective metal is coated with another protective metal. Needless to say, none of this affects anyone's enjoyment of the stories.

The resistance and durability of iron can be greatly increased by adding various materials to it in the molten state. Small amounts of carbon produce various kinds of "ordinary" steel:

| *Percentage of carbon* | *Uses* |
| --- | --- |
| 0.05–0.2 | wire, nails, pipe, stamping sheets |
| 0.2 –0.6 | rails, structural steel |
| 0.6 –1.5 | knives, springs, razor blades |

For the "special" steels, chromium may be added for resistance to strain (car axles), manganese for resistance to wear (railroad rails), nickel for corrosion resistance (gears, cutlery), and so forth—according to the needs of the purchaser.

A typical "stainless steel" for high-quality cutlery contains 1 percent carbon and 18 percent chromium. A typical steel for aircraft or railroad cars contains 0.2 percent carbon, 8 percent nickel, and 18 percent chromium.

*Cobalt* (*Co*) and *Nickel* (*Ni*). Although these two metals were isolated in the eighteenth century, their names hint at chemistry's older and darker traditions.

Cobalt comes from the German *Kobold,* a gnome or goblin reputed to live inside mountains. Nickel is short for *Kupfernickel,* copper of the Old Nick, or Devil's copper. Early miners, unable to extract copper from cobalt and nickel ores, attributed their failures to the malicious interference of gnomes or the Devil.

Cobalt is used to color glass blue and to produce highly magnetic alloys. Alnico (an alloy of iron plus Al, Ni, and Co) can lift 4,000 times its own weight in iron.

The United States "nickel," or five-cent piece, is 25 percent nickel and 75 percent copper.

*Mercury* (*Hg*). Chemically inactive, mercury conducts electricity well and is used in silent electrical switches. When mercury vapor conducts electricity it also emits light and ultraviolet rays; fluorescent lamps and sunlamps make use of these properties.

Mercury dissolves other metals (except iron and platinum) to form mercury alloys called "amalgams." This is one method of separating gold from its impurities.

Mercury is poisonous but also insoluble, which limits its ability to penetrate living cells. Repeated exposures are highly dangerous, however. In nature, mercury may form toxic and readily absorbed compounds such as methyl mercury ($CH_3Hg$).

Mercuric chloride ($HgCl_2$) is a useful antiseptic and disinfectant (diluted 1 : 1,000). It denatures protein; when taken internally it blocks kidney function and is rapidly fatal. Milk and egg white are effective antidotes if given soon enough; the protein they contain ties up the mercury and prevents it from being absorbed.

*Aluminum* (*Al*). Although this is the most abundant metal on the earth's surface, it is so active chemically it is never found in the free state. It was finally isolated in the nineteenth century, but remained so rare that Napoleon III had a tableware set made of it and used it at major state functions.

An American, Charles Martin Hall (1863–1914), in 1886 invented an electrolytic process for purifying aluminum (still used today) which quickly brought down the price per pound from $100 to 15 cents. The last emperor of the French did not live to see this circumstance.

In spite of its high reactivity, aluminum resists corrosion. A tough, thin, transparent film of aluminum oxide ($Al_2O_3$) forms on the surface of the metal when it is exposed to air, protecting it from further chemical change. It weighs less than three times the same volume of water (platinum weighs 21 times the same volume of water). This lightness has made aluminum indispensable to the aircraft, automobile, and building trades.

*Magnesium* (*Mg*). Even lighter and more reactive than aluminum, magnesium was produced only on a small scale before the Second World War. Then Dow Chemical Company developed a cheap and ingenious process for extracting it from sea water. As a result, magnesium production has greatly expanded, as has Dow Chemical.

Besides its use in light-weight alloys, magnesium is a vital part of flares, fireworks, and incendiary bombs, since it burns vigorously with a very bright light. It also reacts with boiling water (slowly) and with steam (rapidly) to form magnesium oxide ($MgO$) and the inflammable gas hydrogen ($H_2$).

The metals most important to Western civilization at the present time are two ancient ones—iron and copper—and two modern ones—aluminum and magnesium.

# 4. The Chemical Revolution Begins

*I am a great lover of chymical experiments.*
——Robert Boyle, *The Sceptical Chymist*

The year 1750 is often singled out as marking the start of modern chemistry. Did chemists all over the world throw off alchemical robes that year and put on laboratory coats? No. But by that time most chemists found it more rewarding to learn new facts on their own by experimentation than to puzzle over an ancient (and by now somewhat garbled) alchemical tradition.

As usually happens, the change had been initiated much earlier by an unappreciated pioneer, in this case Johann Baptista van Helmont (1579–1644).

A member of a noble family in Brussels, van Helmont was well educated, well traveled, and well read. After an early enthusiasm for the writings of the medieval mystics (Thomas à Kempis, Meister Eckhart, Johannes Tauler, the Qabbālāh, and the *Disquisitionum Magicarum* of the Jesuit scholar Martin Del Rio) and for medical works, ancient and modern, he gave away his books, feeling they had not really taught him anything of value. At the age of thirty he obtained an M.D. degree, married, and settled down—not to practice medicine, however, but to devote himself to chemical experimentation.

Some of the mysticism lingered on. Van Helmont once had a vivid and overwhelming vision in which he saw his own soul as a resplendent crystal. In his writings he casually mentions that a stranger gave him a red powder which enabled him to turn mercury into 2000 times its weight in gold. He believed in spontaneous generation—that incubating wheat for three weeks would produce mice.

But at the same time he was making original and accurate observations on gases, plant growth, digestion, fermentation, and chemical change. He described the causes of bronchial asthma (including frustration and repression). He stated that matter is indestructible—that when metal is dissolved in acid it is not destroyed but can be completely recovered. He systematically attacked the Aristotelian theory of four elements. He recognized that the gas produced by beer fermentation (carbon dioxide, $CO_2$) was the same gas that was produced by burning wood or alcohol or by treating sea shells with vinegar. He had no idea what the gas actually *was,* but he knew that he could produce the same one by three different methods.

His equipment kept blowing up when he generated gases. Even the strongest possible vessel, he complained, "straightway leapeth asunder into broken pieces." This led to his inventing the term "gas," which he derived from the Greek "chaos."

Van Helmont's "tree experiment" is an example of his patient curiosity. He wanted to know where the extra bulk came from when a tree increased in size. He planted a 5-pound willow in 200 pounds of soil in a tub. For five years he added nothing to the tub except distilled water and rain water. Then he weighed the soil again and found that it had only lost two ounces. But the willow now weighed 169 pounds. He concluded that the 164 pounds the willow had gained came from water. This is largely true, though what he did not know was that $CO_2$ from the air had joined with some of the water to make carbohydrates—the process now called photosynthesis.

So much independent thought aroused the suspicions of the Inquisition, which brought charges against van Helmont, confiscated some of his papers, and kept him confined to his house for the last ten years of his life. When plague swept through the country, his family refused to leave without him; two of his sons died of the disease. Two years after his own death, van Helmont was acquitted.

A surviving son, Francis Mercurius van Helmont (1614–1699), also studied medicine and chemistry. He seems to have been bright but lacking in perseverance. Part of his youth was spent leading an "irregular life" among gypsies, whatever that entailed. In accordance with his father's deathbed request, he collected 118 of his father's papers, which were published in 1648 under the title *Ortus Medicinae* (Origins of Medicine).

The book strongly influenced later chemists (Boerhaave committed whole sections of it to memory) and was widely translated. Publishers of the English edition (1662) evidently found the title lacking in impact, so they embellished it as follows:

"*Oriatrike or Physick Refined,* The Common Errors therein Refuted, And the whole Art Reformed & Rectified: Being a New Rise and Progress of Phylosophy and Medicine, for the Destruction of Diseases and Prolongation of Life. Written by that most Learned, Famous, Profound, and Acute Phylosopher and Chymical Physitian, John Baptista Van Helmont, and now faithfully rendered into English, in tendency to a common good and the increase of True Science."

Van Helmont's works were closely studied by the English scientist Robert Boyle (1626–1691), who is—to his fellow countrymen, at least—"the father of chemistry."

Like many scientists of the seventeenth and eighteenth centuries, Boyle was an inspired and largely self-taught amateur. He distrusted Aristotle as a student of nature, though not as a moralist or literary critic. Totally committed to experimentation and blessed with a wealthy sister, he turned himself into a kind of research factory. There is something quite modern in Boyle's practice of having assistants perform a number of experiments simultaneously, while he dictated paper after paper to various secretaries.

Using an air pump, he studied the pressure / volume relationships of gases (Boyle's Law). He invented techniques for qualitative analysis, including flame tests, indicators, precipitates, specific gravity, odor and solubility tests. He studied color, fluids, the relation of science to religion. In a famous definition in *The Sceptical Chymist* (1661), he said that chemical elements were "certain Primitive and Simple, or perfectly unmingled bodies . . . not being made of any other bodies. . . ." While this has a surprisingly modern sound, Boyle, although he systematically showed that the Aristotelian elements (and substances like oil, salt, and phlegm) were certainly not elements, was far from understanding elements as they are known today and gives no examples of them.

One of Boyle's greatest contributions was his realization that chemistry was a part of "natural philosophy," not a weird mystical art. He also emphasized that chemistry deserved to be studied for its own sake, not merely as a handmaiden to medicine.

Boyle played a key role in the formation of the Royal Society of London. Students of national character can brood over the fact that in the seventeenth century the English formed a scientific organization (the Royal Society), while the French formed a literary organization (the Académie Française). Yet neither

French science nor English literature seems to have suffered thereby.

*Nullius in verba* (On the word of no one) was chosen as the Royal Society's motto. The way to truth was through experiment, not disputation. Like the Académie Française, the Royal Society was eager to reform language and style. Bishop Thomas Sprat reported in his *History of the Royal Society of London* (1667) that its members promised "to reject all amplifications, digressions, and swellings of style: to return back to the primitive purity and shortnes . . . a close, naked, natural way of speaking . . . preferring the language of Artizans, Countrymen, and Merchants before that of Wits and Scholars."

One problem particularly vexed chemists at this time: What actually happens when something burns? Not even Boyle was much help here. Noting that metals often increase in weight when heated, he suggested that "igneous corpuscles" from the fire were absorbed by the metal.

The self-taught German chemist Johann Becher (1635–1682?) had a different idea. Building on medieval expansions of alchemical theory, he claimed there were three different kinds of "earth"; one of these, *terra pinguis,* was inflammable. When a substance contained *terra pinguis,* it could burn and in the process lose its *terra pinguis*. Alchemically oriented and something of an entrepreneur, Becher persuaded the city of Haarlem to buy his secret process for turning silver into gold.

At the University of Halle, Georg Ernst Stahl (1660–1734) took up Becher's theory. Stahl taught medicine, physiology, dietetics, chemistry, botany, and anatomy, making each lecture as difficult as possible. In 1716 he became physician to Frederick William I (father of Frederick the Great) and apparently found the frigid Berlin court exactly suited to his temperament.

Harsh, sarcastic, and proud, Stahl either ignored the work of other chemists or recast their ideas in his own words. He changed Becher's *terra pinguis* into *phlogiston* (a Greek word meaning "inflammable"), and the notorious phlogiston theory was born. Here is a specimen of Stahl's scientific writing as translated by J. R. Partington in his *History of Chemistry* (1961):

"Briefly, in the act of composition, as an instrument there intervenes and is most potent, fire, flaming, fervid, hot; but in the very substance of the compound there intervenes, as an ingredient, as it is commonly called, as a material principle and as a constituent of the whole compound the material and principle of fire, not fire itself. This I was the first to call phlogiston."

This was a far cry from the Royal Society's return to "primitive purity and shortness." Nor does Stahl get any clearer, as he goes on. Plants extract phlogiston from the air. Soot is almost pure phlogiston. Phlogiston produces colors, odors, and metallic properties, yet is so subtle it cannot be measured or weighed. Much of the theory's success came from the fact that phlogiston was never precisely defined: it could be light, heat, coal dust, or whatever was needed to explain a given set of facts.

Behind the apparatus of pedantry, the core of the phlogiston theory was childishly simple. Things burn because they are burnable. The same reasoning occurs in the last scene of Molière's *Malade Imaginaire* (1673), where the hero explains to a committee of doctors that opium causes sleep (*facit dormire*) because it contains a Sleep Principle (*virtus dormativa*). A single sentence by Sir Isaac Newton (1642–1727) in his *Opticks* (1704) exposes the basic fallacy of the phlogiston theory:

"To tell us that every species of Thing is endowed with an occult specifick Quality by which it acts and produces manifest Effects, is to tell us nothing."

But few listened. The phlogiston theory became almost universally accepted. The science which had broken free of alchemy now became enmeshed in a theory far more treacherous than the transmutation of metals. A scientific hypothesis must be capable of being refuted when it makes a false prediction. But for nearly a hundred years the phlogiston theory made false predictions, and each time the theory was merely softened and twisted to accommodate the conflicting data. If nothing disproves a hypothesis, the hypothesis explains nothing.

A supercool, subtle, independent mind was needed to get chemistry out of the mire. Such a mind belonged to Antoine Laurent Lavoisier.

# 5. Lavoisier

*He discovered no new body—no new property—no natural phenomenon previously unknown. . . . His merit, his immortal glory consisted in this—that he infused into the body of science a new spirit.*
——Justus von Liebig, *Chemische Briefe*

Son of a rich Parisian lawyer, Antoine Lavoisier (1743–1794) was first trained for a legal career. But after receiving his Bachelor's degree and Licenciate, he turned to science. The king gave him a gold medal for his essay on city illumination; at twenty-five, he was elected a member of the Académie des Sciences. A few years later, he married the fourteen-year-old daughter of a business associate. Jacques Louis David's famous portrait of the couple can be seen at Rockefeller University in New York, Lavoisier suave and elegant in a powdered wig, his wife leaning over him solicitously and looking rather like Marie Antoinette.

Madame Lavoisier became the kind of wife chemists dream

about. She entertained her husband's celebrated visitors, worked with him in the laboratory, kept his notes, made sketches for his articles, and translated the works of rival English chemists as fast as they were published.

After some early experiments showing that diamonds will burn (which was already known) and refuting an old notion that water can turn into earth (already refuted), Lavoisier began a series of experiments on combustion. Borrowing a sensitive analytical balance from the French Mint, he weighed pieces of sulfur and phosphorus before and after heating. Without question, each substance gained weight after heating. According to the phlogiston theory, they should have lost weight, since phlogiston was being driven off.

In a 1772 notebook he wrote down plans for more experiments which he felt were "destined to bring about a revolution in chemistry." Once he had all the facts, he would link them up and "form a theory." Unlike many of his colleagues, he did not try to make the facts fit a pre-existing theory. He let the facts suggest a theory, then he tested it. Much of his greatness lies in this approach.

On the other hand, he had no compunction about publishing other people's experiments as his own or predating his papers by several years so that he could claim to have made a discovery before the actual discoverer.

In 1774 the English chemist Joseph Priestley (1733–1804) visited Paris and described to Lavoisier his recent discovery of oxygen (or "dephlogistonated air" as he called it). Lavoisier immediately repeated these experiments and published the results, without giving Priestley the slightest credit. (Lavoisier's occasional uses of innuendo or omission of a key fact to give an impression contrary to the truth are meticulously detailed by J. R. Partington in *A History of Chemistry;* see Suggestions for Further Reading.)

Lavoisier was appointed *Régisseur des Poudres* in 1776, with the task of improving French gunpowder, then the poorest in Europe. He and his wife moved into the luxurious quarters and laboratories of the Royal Arsenal, which quickly became a glittering social center. Benjamin Franklin and Thomas Jefferson were among his visitors. Yet in spite of these distractions, Lavoisier completed the gunpowder assignment. A first-rate product was ready for Napoleon's military adventures.

At this time, poor women used to collect wood ashes from the streets and sell them to saltpeter manufacturers for 2 sous and 6 deniers a bushel. Lavoisier analyzed the ashes and reported they were only worth 1 sou and 6 deniers a bushel. The impact of this study on the women has not been recorded.

By 1776 Lavoisier felt ready to make a direct attack on the phlogiston theory. He wrote that "l'air dèphlogistique de M. Prisley" (when he finally mentions Priestley, he misspells his name) was a normal component of the air. When elements like phosphorus were heated, they combined with some of this gas and became heavier. He called the gas *oxygène,* "acid former."

$$\underset{\text{(phosphorus)}}{4P} + \underset{\text{(oxygen)}}{5O_2} \rightarrow \underset{\text{(phosphorus pentoxide)}}{2P_2O_5}$$

The oxide of phosphorus is a white powder. But when the element carbon is burned—that is, combines with oxygen—the resulting oxide is a colorless gas:

$$\underset{\text{(carbon)}}{C} + \underset{\text{(oxygen)}}{O_2} \rightarrow \underset{\text{(carbon dioxide)}}{CO_2\uparrow}$$

The oxides produced by Lavoisier from nonmetallic elements had something in common; whether they were solids or gases, they reacted with water to form acids. Phosphorus pentoxide formed phosphoric acid, carbon dioxide formed carbonic acid, and so on. This was why Lavoisier named the new gas "acid former."

There was one annoying exception to Lavoisier's theory.

The gas hydrogen ($H_2$), called "inflammable air" by the phlogistonists, was certainly not a metal and it certainly burned (combined with oxygen), but try as he would Lavoisier could never find any trace of acid after the reaction.

The answer was supplied by Henry Cavendish (1731–1810), the eccentric but painstaking English chemist who had discovered hydrogen in the first place. Cavendish found in 1781 that when he "exploded" two parts of hydrogen with one part of oxygen, droplets of "dew" formed on the walls of the container. Moreover, the weight of the water formed exactly equaled the weight of the two gases used up in the reaction.

This bombshell of an experiment was still unpublished in 1783 when a friend of Cavendish, Charles Blagden, visited Paris and told Lavoisier about it. Lavoisier hastily (and rather crudely) repeated the Cavendish experiment and reported the results to the Académie the very next day. In his written report, Lavoisier dismissed the work of Cavendish as not being quantitative (when it was), described his own experiment as quantitative (when it was not), and climaxed his misrepresentation by letting his report appear under the date 1781.

Cavendish was no fighter, but Blagden, in a devastating letter published in *Crell's Annalen 1786* (a widely read clearinghouse of scientific information), described what had happened and stated bluntly that Lavoisier had "discovered nothing beyond what had been pointed out to him as having previously been done and demonstrated in England."

Lavoisier responded with equivocation and legalistic quibbling for several years, then suddenly backed down. He admitted that he had only "confirmed" the synthesis-of-water experiments of Cavendish. The great mathematician-astronomer Pierre Simon de Laplace (1749–1827), who had worked with Lavoisier on a number of experiments, seems to have used his influence on him to bring an end to the degrading affair. Son of a Normandy farmer, Laplace had no taste for Parisian sophistries or sly one-upmanship.

Lavoisier renamed "inflammable air" *hydrogène,* "water former." He was able to run the Cavendish experiment backward by passing steam through a heated gun barrel. The water broke down into hydrogen and oxygen:

$$2H_2O \xrightarrow[\Delta]{} 2H_2\uparrow + O_2\uparrow$$

The free oxygen immediately reacted with the iron of the gun barrel to form rust:

$$3O_2 + 4Fe \rightarrow \underset{\text{(ferric oxide)}}{2Fe_2O_3}$$

A neglected figure among the "gas chemists" of the eighteenth century was the Swedish pharmacist Carl Scheele (1742–1786). Working in relative poverty and isolation, he was the first to discover oxygen (shortly before Priestley did) and the first to realize that air is not a single element but a mixture of gases. He called the two major gases *elds luft* (fire air) and *skamd luft* (spoiled air). "Fire air" was oxygen. "Spoiled air" was called *azote* (no life) by Lavoisier because it was incapable of supporting life; another Frenchman, Jean Chaptal, proposed the name *nitrogène* (niter former) from the old term for potassium nitrate ($KNO_3$). Nitrogen (N) is the name now used everywhere except in France, where azote (Az) still prevails.

Scheele estimated that ordinary air consists of 1 part oxygen to 3 to 4 parts nitrogen. Using expensive equipment and decimals to the eighth place, Lavoisier got still more variable results: 1 part oxygen to 4 to 6 parts nitrogen. It was Priestley who obtained the correct ratio of 1 part oxygen to 4 parts nitrogen. In other words, the air we breathe is roughly 20 percent oxygen and 80 percent nitrogen.

One reason why alchemy had achieved so little, after fifteen centuries of effort, was the terrible confusion in the terminology.

Each new generation had to learn all over again what the previous generation had learned and not disclosed. Alchemical books were little help. In a moment of candor, the Italian alchemist Petrus Bonus wrote in *Pretiosa Margarita Novella* (1330) that all the "secrets of alchemy" that had ever been published could be expressed in six to twelve lines.

"Occult" means "hidden." As an occult science, alchemy went to great lengths to hide its secrets. Analogy, symbolism, secret names, codes, omission of vital facts were combined with a maddeningly evasive, long-winded style. Added to the deliberate obscurity were accidental errors. Translators made mistakes. In one treatise, the Arabic word *dârsini* (cinnamon) emerged as "arsenic" in a list of useful medicines. (This must have thinned out many a doctor's clientele.) *Nitrum* (soda) and *vitrum* (glass) were often confused. Copyists made mistakes, omitted or added words, introduced their own comments without identifying them as such.

Lavoisier undertook to reform chemical terminology. Working with three other chemists, Antoine François de Fourcroy, Louis Bernard Guyton de Morveau, and Claude Louis Berthollet, he produced the *Méthode de Nomenclature Chimique* in 1787. Terms like "butter of Antimony," "pomphlix," "materia perlata of Kerkringius," and "phogadenic water" were ruthlessly eliminated. Terms that agreed with the new chemical discoveries were introduced. One-third of the book consisted of dictionaries of synonyms of the old and the new terminology.

The book was not an overnight success. A committee of the Académie des Sciences in its *Rapport sur la Nouvelle Nomenclature* (1787) deplored the prospect of giving up the phlogistonistic terminology by means of which chemistry had expressed itself *"avec une merveilleuse clarté"* for half a century. The Irish chemist Richard Kirwan (1733–1812) took exception to the term "oxide." "How impossible! In pronouncing this word it cannot be distinguished from 'hide of an ox.' Why not use

Oxat?" he wrote in the *Transactions of the Royal Irish Academy* (1803). But the long-range influence of the *Méthode* was immense.

The wrangle over who deserved credit for the various "gas" discoveries continued. Lavoisier, Priestley, Scheele, Cavendish, the Scottish inventor James Watt (1736–1819), the French chemist Pierre Bayen (1725–1798), and a number of other chemists were involved. Certainly the others did the experiments before Lavoisier. Certainly Lavoisier was incredibly vain and totally unscrupulous in his efforts to establish his own "priority." Nevertheless, he was the only one who correctly interpreted the results of these experiments, the only one who understood and explained their true significance.

Lavoisier further consolidated his position by publishing the first textbook of modern chemistry, his *Traité Elémentaire de Chimie* (1789). Simply and lucidly written, it became an immediate bestseller and was translated into English, German, Dutch, Italian, and Spanish. In the *Traité* Lavoisier defines an element as anything which chemical analysis cannot break down into something simpler: he lists 33 of these. He shows that combustion is simply the union of a substance with oxygen. No mysterious phlogiston theory need be invoked to explain it.

Lavoisier's *Traité* was a superb device for converting the younger chemists. The older ones did not read it. "All young people adopt the new doctrine," Lavoisier noted with satisfaction, "and from this I conclude that the revolution in chemistry is accomplished."

Unfortunately, another revolution was just beginning. The *Traité* was published in 1789, the first year of the French Revolution.

In his youth, Lavoisier had bought his way into the Ferme-Générale, a corporation which leased from the government the right to collect taxes. Profits were limited only by the corporation's scruples about extorting money from the poor. Such scruples did not exist.

Had Lavoisier isolated himself in the laboratory, so that he was unaware how his income was being obtained? On the contrary, he had taken an active administrative role in the Ferme-Générale throughout his career. Not until the Revolution had actually begun did he suggest that some reforms in the taxation system might be desirable.

It was too late, of course. The hatred generated by years of abuse now demanded satisfaction. The revolutionary leader Jean Paul Marat took pains to direct that hatred particularly against Lavoisier. Despite poverty and ill-health, Marat had once hoped for a scientific career. It was Lavoisier who had criticized Marat's *Recherches Physiques sur la Feu* (1780) and helped to block his election to the Académie.

The Ferme-Générale was abolished in 1791, and an investigation into its methods and finances was begun. Delays and prevarications enraged the extremists. In 1793 all former tax collectors, including Lavoisier, were arrested.

Antoine Dupin, the prosecutor, was disposed to transfer Lavoisier out of Paris and beyond the reach of the Terror, provided Madame Lavoisier herself made the request. She did so, but more as an infuriated common scold than as a humble suppliant. (This was the first hint of Madame Lavoisier's temper, which reached its fullest development in her second marriage.) Dupin changed his mind.

The tax collectors were interrogated May 7, 1794. They were given copies of the charges against them at one o'clock the next morning. At ten o'clock they were brought before the Revolutionary Tribunal and allowed fifteen minutes to prepare their defense. Not surprisingly, all were found guilty. They were guillotined a few hours later.

That same day, Priestley was on a ship bound for America. His revolutionary views had made life in England difficult; a mob actually sacked his house. In science, however, Priestley's views remained as reactionary as ever. Ten years later, on a farm in Pennsylvania, he was writing his last scientific paper, *The*

*Doctrine of Phlogiston Established and that of the Composition of Water Refuted.*

Perhaps the light of Lavoisier's intelligence was too cold to inspire genuine friendship. Although a number of his colleagues in chemistry had attained influential positions in the revolutionary government, none of them made the slightest effort to save him; in fact, several of them, including Fourcroy and Guyton de Morveau, worked against him.

Even David, who had painted his portrait, turned against Lavoisier. As a friend of the powerful Maximilien Robespierre, David became a deputy. He voted for the execution of Louis XVI, organized revolutionary spectacles, painted the stunning portrait of the assassinated Marat, and succeeded in abolishing the Académie des Sciences.

Only the mathematician Joseph Louis Lagrange (1736–1813) gave Lavoisier a suitable epitaph. He remarked to his fellow mathematician, Jean Baptiste Joseph Delambre, the day after Lavoisier's execution: "It took but a moment to cut off that head; perhaps a hundred years will be required to produce another like it."

# 6. The Earth's Atmosphere

In Lavoisier's opinion, air was a mixture of two gases:

oxygen (supports fire and life), 20 percent;
nitrogen (does not support fire or life), 80 percent.

When carbon burned—whether in fire or in a living organism—it combined with oxygen and carbon dioxide was released. So air also contained traces of carbon dioxide. (Just before his execution, Lavoisier experimented on the Army physician Armand Séguin, measuring the oxygen he took in and the carbon dioxide he gave off.)

The pro-phlogiston forces rejected Lavoisier's theory. They clung to the Aristotelian idea that air was a single element. Whether or not it supported fire depended on how much phlogiston it contained. Cavendish found it impossible to believe that air was a mixture of gases, "as Mr. Scheele and La Voisier suppose."

Yet it was Cavendish who in 1785 detected the presence of

a fourth gas in air. After treating an air sample with a variety of chemical agents, he was left with one small bubble ("not more than $1/120$th part of the whole") which would not react with anything. (This gas was not formally identified until a century later.)

Thus by the end of the eighteenth century the "pneumatic chemists," as scholarly historians like to call them, had accounted for 99.98 percent of the air we breathe, though not all of them realized that they had done so. Here is a modern analysis of air for the four major components:

| *Gas* | *Formula* | *Percent by volume (dry air at sea level)* |
|---|---|---|
| nitrogen | $N_2$ | 78.08 |
| oxygen | $O_2$ | 20.94 |
| argon | Ar | 0.93 |
| carbon dioxide | $CO_2$ | 0.03 |

Sensitive analytical methods have also shown the presence of four inactive gases related to argon, as well as traces of hydrogen and ozone:

| *Gas* | *Formula* | *Percent by volume (dry air at sea level)* |
|---|---|---|
| neon | Ne | 0.0018 |
| helium | He | 0.0005 |
| krypton | Kr | 0.0001 |
| xenon | Xe | 0.000008 |
| hydrogen | $H_2$ | 0.00005 |
| ozone | $O_3$ | 0.000002 |

Water vapor in the air varies from 7 percent at equatorial coasts to almost zero over deserts, the Arctic, elevations of 6 to 12 miles above sea level. There may also be traces of methane ($CH_4$), nitrogen and sulfur oxides, and the radioactive gas radon (Rn) from the decay of radium. Pollution adds a variety of compounds to the air, many not even identified. Winds blowing over the earth's surface slowly wear rocks away and carry dust into the air, along with salt particles from the oceans.

This transparent mixture of gases is held to earth by gravity. It thins out gradually and blends into outer space, so it is hard to set an outer boundary. But half of the atmosphere's total weight is found within 4 miles of the earth's surface.

The atmospheres of other planets can be determined by spectroscopic analysis. They are not inviting. Beyond Mars, hydrogen and two compounds of hydrogen, methane ($CH_4$) and ammonia ($NH_3$), are the dominant gases. Nitrogen and carbon dioxide prevail on Venus and Mars. None of these gases support animal life.

| *Planet* | *Atmospheric gases* |
|---|---|
| Mercury | none |
| Venus | $N_2$, $CO_2$ |
| Mars | $N_2$, $CO_2$ |
| Jupiter | $H_2$, He, $CH_4$, $NH_3$ |
| Saturn | $H_2$, He, $CH_4$, $NH_3$ |
| Uranus | $H_2$, He, $CH_4$, $N_2$ |
| Neptune | $H_2$, He, $CH_4$, $N_2$ |
| Pluto | unknown |

Applying the same method of analysis to as much of the universe as they can see telescopically, scientists found that similar chemical conditions prevail in the rest of our galaxy as well as in other galaxies. If the abundance of nitrogen in the universe is taken as 1, other elements occur in these proportions:

| *Element* | *Relative abundance (N=1)* |
|---|---|
| hydrogen | 10,000 |
| helium | 1,600 |
| oxygen | 10 |
| carbon | 4 |
| nitrogen | 1 |

The earth's atmosphere is significantly different from the prevailing chemistry of the universe. When first formed, the earth too must have had a hydrogen-rich atmosphere, but it now contains free oxygen and virtually no free hydrogen. The processes by which this happened have been reproduced experimentally. No doubt they have occurred and will occur again in other parts of the cosmos.

But at this particular moment in time we cannot detect (let alone reach) any other place in the universe which could even begin to support our kind of life. The thin envelope of gases around earth—which we pollute so casually—is our only home.

The nature of these atmospheric gases is described in the next five chapters.

# 7. Oxygen

Many chemical elements are necessary for life, and it is foolish to say that one element is "more essential" than another. Yet oxygen is so basic to life that it deserves to be placed at the top of the list. Every moment from conception to death man depends on this element; he cannot survive more than a few minutes without it.

Oxygen is colorless and odorless, with nothing to distinguish it from the other atmospheric gases or from air itself. It can be distinguished by what it does, however. A lighted match will go out in a bottle of carbon dioxide or nitrogen; it will cause an explosive flash in hydrogen; it will burn more brightly in a bottle of oxygen.

At room temperature, water can hold 3 percent dissolved oxygen. The hotter the water, the lower the percent of dissolved oxygen it contains. Boiled water tastes "flat," because the dissolved oxygen has been driven off. Fish are usually more active in cold water, since it holds more oxygen. Artificially raising the temperature and reducing the oxygen in natural waters (thermal pollution) may make fish sluggish or even kill them.

On earth—and nowhere else—oxygen is the most abundant of the elements. Free oxygen makes up 20 percent of the atmosphere. Combined oxygen makes up 45 percent by weight of the earth's minerals, 90 percent of water, and 67 percent of the human body.

An analysis of the earth's surface (including oceans and the air) shows the following values:

| *Element* | *Symbol* | *Percent by weight* |
|---|---|---|
| oxygen | O | 49 |
| silicon | Si | 26 |
| aluminum | Al | 7.5 |
| iron | Fe | 4.7 |
| calcium | Ca | 3.4 |
| sodium | Na | 2.6 |
| potassium | K | 2.4 |
| magnesium | Mg | 1.9 |
| hydrogen | H | 0.88 |
| titanium | Ti | 0.58 |
| chlorine | Cl | 0.19 |
| carbon | C | 0.09 |
| all other elements | | 0.76 |

In the laboratory, oxygen is prepared by decomposing an oxygen-containing compound. Priestley simply heated mercuric oxide ($HgO$), a bright orange powder popular with alchemists:

$$2HgO \xrightarrow[\Delta]{} 2Hg + O_2\uparrow$$

Oxygen and other gases are produced commercially by liquefying air. Each liquefied gas in the mixture has a different boiling point, so as the temperature is slowly raised, one gas after another will evaporate. They are collected separately and stored in steel cylinders.

| Gas | Boiling point (degrees Fahrenheit) |
|---|---|
| oxygen | −297.4 |
| argon | −302.6 |
| nitrogen | −320.5 |
| neon | −410.9 |
| hydrogen | −422.9 |
| helium | −452.1 |

## Oxides

When oxygen combines with other substances the process is called oxidation and the products usually include oxides. When an oxidation is rapid, there is combustion, or fire. But many oxidations take place slowly, without flame: the rusting of iron, the drying of oil-base paints, the browning of apples.

The oxides of nonmetals usually form acids with water:

$$\underset{\text{(sulfur trioxide)}}{SO_3} + H_2O \rightarrow \underset{\text{(sulfuric acid)}}{H_2SO_4}$$

This fact misled Lavoisier into thinking that all acids must contain oxygen. He was unacquainted with hydrochloric acid (HCl), a strong acid containing no oxygen.

The oxides of metals usually form alkaline compounds with water:

$$\underset{\text{(aluminum trioxide)}}{2Al_2O_3} + 6HOH \rightarrow \underset{\text{(aluminum hydroxide)}}{4Al(OH)_3}$$

Remedies for "acid stomach" often contain this hydroxide or others like it. Milk of magnesia is magnesium oxide (MgO) which slowly dissolves in water to form magnesium hydroxide:

$$MgO + HOH \rightarrow Mg(OH)_2$$

Because they are so weakly alkaline, these hydroxides can be safely used as remedies. Other hydroxides are far too strong and corrosive. Potassium hydroxide (KOH) and sodium hydroxide (NaOH, lye) are used to clear stopped drains.

The most abundant oxide on the earth's surface is water ($H_2O = HOH$). Silicon dioxide ($SiO_2$) is likewise extremely common, being found as sand, quartz, amethyst, agate, opal, onyx, flint, and in glass.

Hydrogen peroxide ($H_2O_2$) is an unstable, colorless liquid which readily breaks down to form water and oxygen:

$$2H_2O_2 \rightarrow 2H_2O + O_2\uparrow$$

It is useful as an antiseptic (3-percent solution), a bleaching agent (for hair, textiles, straw, ivory, fats and oils), and as an ingredient of rocket propellants.

Another oxide, carbon dioxide ($CO_2$), is so important that chapter 8 is entirely devoted to it.

## "Animal Heat"

The early history of man is filled with sun worship and legends involving fire. Even today we maintain Eternal Flames and light candles for worship or for yoga exercises. Songs, poems, and proverbs have always compared human life to a flame. How close is the kinship?

The ancients puzzled over why warm-blooded animals are warm. Plato suggested in *Timaeus* that the heart was the center of an "inner fountain of fire," which it was the function of the lungs to cool. Galen compared a living man to an oil lamp—each burning stored fat as fuel and each extinguished when deprived of air.

After discovering that blood circulates, the English physician William Harvey (1578–1657) concluded that its purpose was to warm and nourish all parts of the body. Lavoisier showed that combustion is a reaction with oxygen; when carbon or carbon compounds burn, the product is $CO_2$. When he discovered that animals take up oxygen as they respire and give off $CO_2$, the similarity was striking indeed:

Combustion:

Organic material (wood) + $O_2 \rightarrow CO_2 + H_2O$ + energy

Respiration:

Organic material (starch) + $O_2 \rightarrow CO_2 + H_2O$ + energy

Now it was only necessary to find out *where* in the animal body this reaction occurred. Lavoisier thought it must be in the lungs, where oxygen in the air contacted food materials dissolved in the blood. Lagrange criticized this on common-sense grounds. If the lungs were several degrees hotter than the rest of the body, the fact would have been noticed long ago. He suggested that oxygen dissolved in the bloodstream and that oxidation occurred in all parts of the body reached by the circulating blood. This was confirmed experimentally by nineteenth-century physiologists. In more precise terms, oxidation occurs in all tissues with oxygen released by the capillaries. The resulting $CO_2$ is carried back in the bloodstream to the lungs and there exhaled.

Though Lavoisier used the expression *"la machine animale,"* warm-blooded animals are not merely heat engines. An engine must have a heat source; to give the same efficiency observed in animals, the heat source would have to be close to the boiling point. Nothing of the kind takes place in the animal body. Something far more subtle is at work.

## Fire and Explosion

Things that burn usually have rather sharply defined "kindling temperatures." These may be reached unintentionally. Grains, newspapers, and oily rags oxidize slowly in air. If they are stored so that the heat cannot escape as fast as it is formed, the kindling temperature will be reached and "spontaneous combustion" will result.

The two strategies for putting out fires are: (1) lower the temperature; (2) cut off the oxygen supply.

Throwing water on a fire tends to lower its temperature;

covering it with dirt or a rug tends to smother it. A carbon-dioxide fire extinguisher does both. Containing liquid $CO_2$ under pressure, it releases the $CO_2$ in the form of a rapidly expanding gas. Just as gases heat when compressed, they cool as they expand. The extinguisher covers the fire with a blast of cold $CO_2$ which both lowers the temperature and separates the fire from the oxygen in the air.

Furniture and fabrics are not damaged by a $CO_2$ extinguisher, but the room must be ventilated afterward; otherwise the user might save himself from the fire only to suffocate from the accumulated $CO_2$.

A fire is a vigorous oxidation. Oxidations can take place even more vigorously in the form of explosions.

It is hard to get a log to burn, but if the log is split into kindling, the task is easier. If the log is reduced to paper-thin shavings, it will burn instantly. The same amount of wood is involved, but shavings put more wood surface in contact with the oxygen of the air, so that the wood can react and burn more easily. Very fine sawdust floating in the air presents even more surface than wood shavings. When one sawdust particle ignites, it kindles all the particles around it, which kindle the particles around them, and so on in an instantaneous chain reaction. There is a sudden burst of heat and expansion of gases: a violent explosion with multiple fires.

Because of this possibility, the presence of flour, coal dust, gasoline fumes, or any inflammable dust or vapor in the air is always extremely dangerous.

## Spoilage and Aging

Reactions with oxygen cause many types of food to spoil. While foil wrapping, cans, and bottles prevent oxygen from reaching the food, they are relatively expensive. Another solution is to add a chemical "anti-oxidant" that will slow down the reactions with oxygen. Some anti-oxidants are harmful; none are nutritionally valuable. They are not there for the consumer's benefit, but to increase the shelf life of the food and save the manufacturer money. They can be found among the mysterious ingredients listed on packaged crackers, breakfast cereals, dog food, even bread.

If life is a slow fire, is it dangerous to speed it up with extra oxygen? Priestley suspected that this might be true. He wrote in *Experiments and Observations on Different Kinds of Air* (1775), "For as a candle burns out much faster in this air [oxygen] than in common air, so we might *live out too fast*."

Patients who have received 80 to 100 percent oxygen for long periods of time show many signs of mental and physical deterioration. On the other hand, mice have been made to live longer than usual in experiments that slowed down their oxidative processes in various ways.

## Ozone

In all the chemical equations given so far, oxygen is shown as $O_2$. This is because oxygen atoms pair up with one another, forming double atoms, or "diatomic molecules." Hydrogen ($H_2$) and nitrogen ($N_2$) do the same thing; inactive gases such as neon (Ne) do not.

Under some conditions (ultraviolet light, electrical discharges, pollution) oxygen forms a triatomic molecule called ozone, from the Greek word meaning "to smell."

$$\underset{\text{(oxygen)}}{3O_2} + \text{energy} \rightarrow \underset{\text{(ozone)}}{2O_3}$$

Ozone is an irritating bluish gas with a pungent, somewhat garlicky odor, often noticeable after a thunderstorm. In traces it can be pleasant and at the turn of the century it was considered healthful; streets—and even towns—were named after it. It is toxic, however, and may even be fatal in concentrations as low as 0.0006 percent.

Ozone is more reactive than ordinary oxygen. It blackens silver and attacks rubber, textiles, and dyes. It is used as a bleaching agent and a disinfectant.

# 8. Carbon Dioxide

Because of the key role it plays in biochemistry, carbon dioxide ($CO_2$) has a significance out of all proportion to its low concentration (0.03 percent) in the atmosphere. It also has some unusual physical properties.

Like air, $CO_2$ is colorless and odorless, but it is 1½ times heavier than air, so that it tends to accumulate in mines, caves, and cellars. Amateur explorers risk suffocation when they charge heedlessly into such areas.

Carbon dioxide liquefies at $-78°$ C. and becomes a solid, "dry ice," at $-109°$ C. Instead of melting, dry ice turns directly into gaseous $CO_2$, entirely skipping the liquid phase. The alchemists called the process "sublimation" and purified sulfur and arsenic that way.

Carbon dioxide in the air is transparent to visible light, but tends to screen out the adjoining wavelengths. As a result, most of the sun's lethal ultraviolet rays fail to reach the earth's surface. Carbon dioxide also produces a so-called greenhouse effect. Warmed by the sun, the earth emits infrared (heat) rays. Carbon dioxide blocks the escape of these rays into outer space, so that the atmosphere warms up, rather like a closed car on a sunny day.

When the $CO_2$ level in the air increases, heat retention also increases, and the atmosphere warms up still more. This is actually happening at the present time. Our large-scale burning of "fossil fuels" (coal and petroleum) has caused slight increases in the air's $CO_2$ content. Glaciers are melting. The earth is, in fact, warming up.

It is estimated that if atmospheric $CO_2$ were to double (increasing to 0.06 percent), tropical climates would prevail throughout the world. Icecaps at the North and South poles would melt, releasing enough water to raise the level of the oceans about 100 feet.

Increased water in the oceans would be able to dissolve more $CO_2$ from the air and thus tend to bring the $CO_2$ level down again. The time lag involved is not known.

## Photosynthesis

In 1771, Priestley discovered what seemed to be the reverse of combustion. A mouse would die if he placed it in a container in which a candle had burned to extinction (using up the oxygen). But if he placed a mint plant in this same chamber for a few days, the plant not only did not die but somehow restored the air so that it became "not at all inconvenient to a mouse." The mint plant had released oxygen.

Priestley failed to notice that light was necessary for this effect. The Dutch physician Jan Ingen-Housz (1730–1799) reported that fact in his *Experiments upon Vegetables* (1779). And the Swiss chemist N. T. de Saussure (1767–1845) concluded in

*Recherches Chimiques sur la Végétation* that plants are somehow able to manufacture plant material out of nothing more than $CO_2$ and water, with oxygen released as a by-product.

This extraordinary process is called photosynthesis (photo, light; synthesis, putting together). Energy for the reaction is provided by the plant's green pigment, chlorophyll.

$$CO_2 + H_2O \xrightarrow[\text{(chlorophyll)}]{\text{light energy}} \underset{\text{(carbohydrates)}}{(CH_2O)x} + O_2$$

This explains why plants and animals sealed up in a glass terrarium survive, just as they have survived sealed up together on this planet. It also explains why schemes for long-range space travel often involve vats of green algae: the passengers provide $CO_2$ for the algae, the algae provide $O_2$ for the passengers.

### Carbon Monoxide

Carbon and carbon compounds usually burn to $CO_2$, but if oxygen is in short supply, carbon monoxide (CO) is formed instead:

$$C + O_2 \rightarrow CO_2 \text{ (carbon dioxide)}$$
$$2C + O_2 \rightarrow 2CO \text{ (carbon monoxide)}$$

Carbon dioxide is not poisonous, but it can kill in so far as it cuts off the oxygen supply. Simple lack of oxygen (anoxia) provokes no alarm reaction (unless, of course, someone is strangling you or holding your head under water); one simply loses consciousness and dies. But any increase in the $CO_2$ content of the blood stimulates a "respiratory center" in the brain and makes one breathe harder. When too much jogging makes you gasp for air, you are not trying to get more oxygen: you are trying to get rid of the extra $CO_2$ your muscles are producing.

Carbon monoxide, on the other hand, is a direct and fast-acting poison. This colorless odorless gas combines irreversibly with the hemoglobin of the red blood cells, so that it is unable to function as an oxygen carrier. Oxygen is unable to reach the tissues, and the blood turns a bright cherry-red.

As little as 0.05 percent CO in the air may cause serious illness and permanent brain damage, while 1 percent will cause death in minutes. A heavy cigarette smoker is estimated to have 5 percent of his hemoglobin tied up with CO; this will not kill him, but neither will it help him run the mile.

Carbon monoxide still contains unused energy, so it can be burned as a fuel, like natural gas:

$$2CO + O_2 \rightarrow 2CO_2 + 135{,}272 \text{ calories}$$

Carbon monoxide produced by industrial processes is reused in this way, since no industry will tolerate the waste of its own resources. For the consumer, however, industry has tolerated the CO-producing automobile engine for over half a century.

# 9. The Gifts of Ammon

Oxygen is a little too active chemically to be lived with comfortably at the 100-percent level. After a day or two it may produce respiratory irritation, headache, chest pains, sudden shifts in mood, even convulsions. But diluted down to 20 percent, oxygen causes none of these problems.

The diluting agent in air is the resolutely inactive gas, nitrogen ($N_2$).

Nitrogen "rusts" no metals. "Antinitridants" do not have to be added to keep it from spoiling food. In fact, some foods and beverages are actually packed in nitrogen to preserve them.

With the help of high temperatures, electric sparks, and catalysts, nitrogen can be induced to react with other elements. Once it has, the compound tends to be unusually potent in some way. Nitrogen compounds include explosives (TNT, nitroglycerine), hormones (adrenalin, insulin), proteins (muscle, hair, claws, enzymes), and drugs (antibiotics, poisons, hallucinogens).

A certain irrationality has crept into the naming of nitrogen compounds. Nitric acid ($HNO_3$), nitrates ($—NO_3$), nitrites ($—NO_2$), all deriving from the parent element, seem reasonable enough.* But then one finds compounds with names stemming from azote, Lavoisier's original name for nitrogen: azides ($—N_3$), azo compounds ($-N{=}N-$), diazonium salts ($ArN{=}NX$), and so on. And finally, where did the names ammonia ($NH_3$), amines ($—NH_2$), amino acids, and amides come from?

They came from Amen, supreme god of the Egyptians from the twelfth dynasty onward. The pharaoh Amenhotep IV (Akhnaton) repudiated this god in the fourteenth century B.C. and tried in vain to interest the people in a simple monotheism. When Alexander the Great (356–313 B.C.) conquered Egypt, he consulted the oracle of Amen and was told that his father was not Philip of Macedon, but the god Amen himself. This idea was not entirely unpleasing to Alexander, so he built a temple to Amen, or Ammon, as the Greeks called him.

Conventional fuels are scarce in the North African desert. At the temple of Ammon, camel dung was burned regularly. Over the years the walls and ceiling of the temple became encrusted with a white crystalline material. The Romans called it *sal ammoniac* (salt of Ammon) and the name continued to be used through the Middle Ages.

The salt was ammonium chloride ($NH_4Cl$). When heated, it gives off a pungent gas, a familiar smell in barns and stables, which was named "ammonia":

$$\underset{}{NH_4Cl} \rightleftharpoons \underset{\text{(ammonia)}}{NH_3\uparrow} + \underset{\text{(hydrogen chloride)}}{HCl\uparrow}$$

In chemistry laboratories, the $NH_3$ and $HCl$ fumes from reagent bottles often combine and form white deposits on windows

* The dash before the formula indicates that the chemical entity does not exist independently but is bound to something else: $K—NO_3$, potassium nitrate, for example.

and glassware, just as they did from camel dung in the temple of Ammon.

## The Great Nitrogen Shortage

Nitrogen is essential for all living things, but only in the combined form. Pure nitrogen gas is no help. A plant can die of an acute nitrogen deficiency while surrounded by the 78 percent nitrogen present in the air.

Flashes of lightning produce nitrogen oxides. In the course of a year all the storms on earth are estimated to form about 40 million tons of nitric oxide (NO). But this has little practical significance. Plants need specific amounts of nitrogen compounds at specific times, and random lightning flashes do not come close to providing these.

Some microorganisms have the extraordinary talent of "fixing" nitrogen: they can take nitrogen from the air and change it into a usable nitrogen compound.

Blue-green algae, one of the oldest and most primitive forms of life, can do this. When the island of Krakatoa was rendered sterile by a volcanic eruption in 1878, blue-green algae were among the first forms of life that returned to the island. They acted as colonizers and helped prepare conditions favorable for higher plants. Blue-green algae may have played a similar role in the early history of life on earth.

In India, rice paddies have been inoculated with blue-green algae as a substitute for nitrogen fertilizers. Certain bacteria also fix nitrogen. Farmers have long known that growing clover, soybeans, alfalfa, or peas for a year often restores the fertility of a "worn-out" field. Bacteria (*Rhizobium*) living in the roots of these plants pull nitrogen from the air and leave it in the soil in the form of usable nitrogen compounds.

Some free-living soil bacteria (*Azotobacter, Clostridium*) fix nitrogen when they have to, but they stop doing so if nitrogen compounds become available. Many species of bacteria release available nitrogen to the soil by breaking down organic matter. And some troublemakers convert available nitrogen back into elemental $N_2$.

"Available nitrogen" means chiefly ammonium salts and nitrates. Being soluble, these are easily washed away by rain or irrigation, so there tends to be a chronic shortage of nitrogen in the soil. Many nineteenth-century scientists viewed this shortage as a formidable threat. Sir William Crookes (1832–1919) told the British Association in 1898 that unless a method was found for fixing nitrogen, "the great Caucasian race will cease to be foremost in the world, and will be squeezed out of existence by races for whom wheaten bread is not the staff of life" (*The Wheat Problem,* 1917).

The problem was solved, happily, by a member of the great Caucasian race. Fritz Haber (1868–1934) developed an ingenious process for combining nitrogen and hydrogen at high temperatures and pressures:

$$N_2 + 3H_2 \xrightarrow[\substack{600\text{ atm.} \\ \text{(catalyst)}}]{500^\circ\text{ C.}} 2NH_3 \text{ (ammonia)}$$

The significance of the Haber process to Germany on the eve of the First World War, and to the rest of the world since then, is hard to exaggerate. Some 20 million tons of $NH_3$ are now manufactured yearly by this process. Haber received the Nobel Prize in 1918. In the 1930s, however, the Third Reich found Haber's Jewishness unforgivable and expelled him—into the waiting arms of Cambridge University.

Pure ammonia gas can be used directly as a fertilizer by pumping it into the ground or dissolving it in irrigation water. Or it can be made into various fertilizer salts. Ammonium nitrate ($NH_4NO_3$) is especially high in usable nitrogen. Like many nitrates, however, it is also explosive. Texas City, Texas, was nearly destroyed in 1947 when a cargo ship filled with ammonium nitrate blew up in the harbor.

## Nitrogen Explosives

After the Chinese invented gunpowder (ninth century), their conquerors the Mongols realized its military value and turned it against the Europeans in 1235. A dozen years later, the Oxford monk-scientist Roger Bacon (1214–1292) published, in code, a formula for gunpowder, adding *"et sic facies tonitruum et coruscationem si scias artificium"* ("and so you make thunder and lightning if you know the trick").

Here is Bacon's formula, along with a modern formula for black gunpowder:

| *Ingredients* | | *Percent in Bacon's gunpowder* | *Percent in black gunpowder* |
|---|---|---|---|
| potassium nitrate | ($KNO_3$) | 40 | 75 |
| carbon (charcoal) | (C) | 30 | 15 |
| sulfur | (S) | 30 | 10 |

The principle is the same in both. Heat easily decomposes the potassium nitrate, releasing oxygen which instantly reacts with the carbon and sulfur. The temperature of the explosion is about 2700° C. The gases formed, $CO_2$ and $SO_2$, expand suddenly outward and do the actual "work" of the explosion.

The nitrate is the key ingredient. By purifying this ingredient, Lavoisier rehabilitated French gunpowder. One of his assistants, Eleuthère Irénée Du Pont de Nemours (1771–1834), carried the technique with him to America and in 1802 began manufacturing high-quality gunpowder in Delaware.

Gunpowder was replaced by "guncotton" in the late 1800s. This was made by fully nitrating cotton with nitric acid. Partially nitrated cotton was the basis of the primitive plastics, collodion and celluloid; while not actually explosive, they were highly inflammable. Motion-picture film was made of these materials until 1948.

Guncotton has the advantage of being smooth-burning and

"smokeless." Other new explosives were made around this time by the same technique: transferring the potentially explosive nitro group to innocent materials by means of nitric acid. Alfred Nobel (1833–1896) treated the colorless sweet syrup glycerine with nitric acid:

$$\begin{array}{l} CH_2OH \\ CH\ OH \\ CH_2OH \end{array} + 3HNO_3 \rightarrow \begin{array}{l} CH_2NO_3 \\ CH\ NO_3 \\ CH_2NO_3 \end{array} + 3H_2O$$

glycerine — glyceryl trinitrate (nitroglycerine)

This yellow oily product does not require a spark to set it off. A good bump will do it. Film melodramas such as Henri-Georges Clouzot's *Wages of Fear* (*Le Salaire de la Peur*, 1952) have vividly impressed the public with this fact—so much so, that banks continue to be robbed and airplanes hijacked by persons waving bottles of water.

By absorbing the nitroglycerine onto wood pulp or diatomaceous earth, Nobel reduced its sensitivity to shock. In this form he sold it as "dynamite." With the fortune he made, Nobel endowed a trust fund for annual awards in peace, physics, chemistry, medicine, and literature. The first awards were made in 1901.

Toluene, a colorless hydrocarbon like benzene ("cleaning fluid"), was nitrated by German chemists just in time for the First World War:

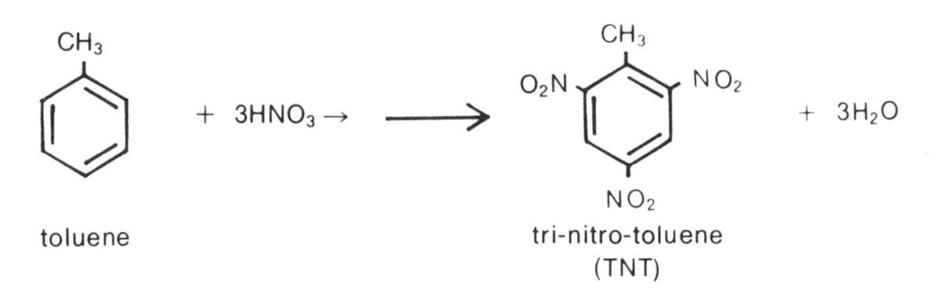


The yellow crystalline product is stable and can only be set off by a detonator.

All these explosives, of course, are minor compared with nuclear weapons.

Except for gunpowder, the explosives just described are "organic" nitrogen compounds, and so are most of the more glamorous nitrogen compounds (dyes, poisons, drugs). But some of the simpler "inorganic" compounds of nitrogen deserve mention:

*Nitrous Oxide* ($N_2O$). When the salt ammonium nitrate is heated (cautiously—remember Texas City), a colorless sweetish gas is given off:

$$NH_4NO_3 \rightarrow N_2O\uparrow + 2H_2O$$

When inhaled, it causes mild hysteria ("laughing gas"), and in larger amounts insensibility to pain and unconsciousness. It was one of the first anesthetics. Vivid hallucinations, often of an erotic nature, may be produced. Dentists long ago learned the wisdom of keeping a third party in the room at such times.

Nitrous oxide became a fad among students when it was first introduced. Descriptions of the psychological effects of $N_2O$-sniffing strongly resemble the accounts of psychedelic enthusiasts.

"I lost all connections with external things; trains of vivid images rapidly passed through my mind. . . . I existed in a world of newly connected and newly modified ideas . . . ," wrote the English chemist Sir Humphry Davy (1778–1829) in his *Researches Chemical and Philosophical, chiefly concerning Nitrous Oxide* (1800).

*Nitrogen Dioxide* ($NO_2$). This brownish poisonous gas is one of the most destructive components of smog. Nitrogen dioxide reacts with water to form nitric acid:

$$3NO_2 + H_2O \rightarrow \underset{\text{(nitric acid)}}{2HNO_3} + NO$$

In smog, this reaction takes place on a small scale in people's lungs and on the surface of their eyeballs.

*Nitric Acid* ($HNO_3$). Powerful and corrosive, this acid is a staple of the chemistry laboratory and industry. Millions of tons are used each year to manufacture explosives, fertilizers, drugs, dyes, movie film, plastics, and countless other synthetic products.

Nitric acid attacks skin proteins, causing yellow burns. Patient murderers have disposed of corpses with nitric acid. Metals are dissolved by nitric acid—except gold and platinum. When hydrochloric acid (HCl) is added, the resulting mixture dissolves even these metals. *Aqua regia* (royal water) was the alchemists' name for this mixture.

*Potassium Nitrate* ($KNO_3$). Besides its use in gunpowder, this salt is a fertilizer and a food preservative. Naturally occurring $KNO_3$ was called "saltpeter" (from *sal petrae,* salt of the rock).

A cherished bit of male folklore holds that adding saltpeter to the diet lessens sexual desire and that this practice is routinely carried out in the Army, prisons, boarding schools, and similar institutions. Pharmacologists are unable to substantiate this. In fact, they can find nothing more significant in saltpeter than a slight tendency to provoke urination—that is, it is a mild diuretic.

*Amyl Nitrite and Nitroglycerine.* Both these compounds have a useful physiological action: they relax smooth (involuntary) muscle. This means they can be taken to relieve the spasms of asthma, coronary attacks, and certain kinds of headaches. The nitroglycerine is taken orally; the amyl nitrite is inhaled. As "poppers," the amyl nitrite vials are alleged to increase the intensity of sexual orgasm.

*Urea.* In the living animal body, protein and other nitrogen-containing compounds are continually being broken down and replaced. The main breakdown product, ammonia, is toxic, but converted into urea it is nontoxic. Most animals excrete their discarded nitrogen in the form of urea.

The alchemists knew about urea. Once they had learned how to distill, it was only a matter of time before they distilled their own urine. When they did, they ended with crystalline urea.

Urea is a useful fertilizer, since it breaks down to form ammonia:

$$H_2N-\underset{\underset{O}{\|}}{C}-NH_2 + H_2O \rightarrow 2NH_3 + CO_2$$

Urea played a pivotal role in the evolution of chemical thought. "Organic chemistry" was applied to the reactions of living organisms early in the nineteeth century. It was assumed that only plants and animals could carry out such reactions. It was also assumed that organic chemical compounds could not be synthesized in the laboratory, but only in the living organism. Then in 1828 the German chemist Friedrich Wöhler (1800–1882) tried to make ammonium cyanate ($NH_4CNO$) from a pair of salts and accidentally synthesized urea instead. "I can make urea without the necessity of a kidney," he wrote to a colleague, "or even of an animal, whether man or dog."

An organic compound had been synthesized in the laboratory, proving that living organisms did not have a monopoly on this kind of synthesis. Organic chemistry had to be redefined and the experimental approaches to it changed accordingly.

# 10. The Inert Gases

The small bubble of unreactive gas found in an air sample by the meticulous Cavendish in 1785 was identified as a new element in 1894 by the Scottish chemist Sir William Ramsay (1852–1916) and the English physicist Lord Rayleigh (1842–1919). They named it argon (from the Greek *argos,* idle, inactive). In the next few years Ramsay and his associates discovered a number of related gases, to which they also gave Greek names:

helium (*helios,* sun)
neon (*neos,* new)
krypton (*kryptos,* hidden)
xenon (*xenos,* strange)

Apparently these exhausted the stock of suitable names from Greek sources, for a sixth gas was named "radon" from the Latin *radium* (ray).

The gases seemed to be completely unreactive. The atoms would not even pair up with each other, like those of oxygen ($O_2$), hydrogen ($H_2$), and nitrogen ($N_2$). But in 1962 a xenon compound was prepared (by sealing the reactants in a nickel vessel and heating it to 400° C.), and later krypton and radon compounds were also made.

A grave semantic crisis faced the chemical world. If these gases were capable of reacting, after all, how could they be called "the inert gases"?

Some chemists now called them "the rare gases." But argon in the air and helium in natural gas deposits are hardly rare; in fact, helium is the second most abundant element in the universe.

Other chemists called them "the noble gases," a name that involves dubious assumptions about the nature of aristocracy. Although the group of "noble metals" was claimed as a precedent and justification, this term is no longer used for the metals.

Actually, inertness does not mean total inactivity, merely a tendency in that direction, so there is nothing wrong with continuing to call the group "the inert gases." Or if you prefer to evade the problem completely, you can call them "the helium-group gases."

Inactivity can be useful. When Hannibal invaded Italy in the third century B.C., Quintus Fabius saved the badly frightened and outnumbered Romans by tactics of "masterly inactivity." The filament of an electric light bulb would quickly burn out in ordinary air, but if the bulb is filled with the inactive gas argon, the filament will last as long as the manufacturer wants it to last.

When an electric current is run through a glass tube filled with one of the inert gases, a brightly colored glow is produced:

| | |
|---|---|
| neon | orange to red |
| helium | pink to white |
| krypton | pale blue |
| xenon | blue to green |

Neon was the first of these gases to be used and it is still the most common. The twentieth century would not look the same without it.

Krypton lamps penetrate fog well, so they are used at airports. Krypton is also produced by nuclear reactors, so it is monitored in the air to detect the unreported operation of such reactors.

Radon, a breakdown product of radium and itself radioactive, has the distinction of being the heaviest of the gases (222 times heavier than hydrogen). Sealed in small capsules, it can be implanted in the body to irradiate cancerous tissue.

Helium, of course, is used for filling balloons and other lighter-than-air vehicles. It is not quite as efficient as hydrogen, but it has the advantage of being noninflammable. The "lifting power" of a gas depends on how much lighter the gas is than air. One thousand cubic feet of air weighs 76 pounds. The same amounts of hydrogen and helium weigh 5 and 10 pounds respectively.

Helium has another useful property: it is not very soluble in water or human blood. In the old days, a deep-sea diver would breathe an air mixture (20 percent oxygen, 80 percent nitrogen) under pressure. The nitrogen readily dissolved in the bloodstream; when the pressure was reduced—that is, when the diver surfaced—the nitrogen bubbled out of the blood, like gas from a freshly opened bottle of beer, causing a painful and dangerous condition known as decompression sickness, or "the bends." By using helium in place of nitrogen, a diver avoids this danger.

The ending "-ium" is supposed to indicate that an element is a metal. The British feel so strongly about this that they call aluminum "aluminium." Yet they named helium, which is certainly not a metal—another proof of how hard it is to be consistent in taxonomic matters.

Naming helium after the sun was quite correct, however. The sun is an enormous nuclear-fusion reactor, where hydrogen is continually transformed into helium:

$$4\,^{1}_{1}H \rightarrow \,^{2}_{4}He + \text{energy} + 2 \text{ positrons}$$

(hydrogen nuclei) (helium nucleus)

This is hardly a conventional chemical reaction, since the temperature at which it occurs is 20 million° F. and since one element is being transformed into another. In so far as human beings depend on the sun for life, they depend on this reaction.

Now that men can make their own nuclear energy, certain problems are involved in having the reaction occur on the earth's surface rather than 93 million miles away.

# 11. The Primeval Element

Of all the elements, hydrogen is the simplest. Its atomic structure can be diagramed as follows:

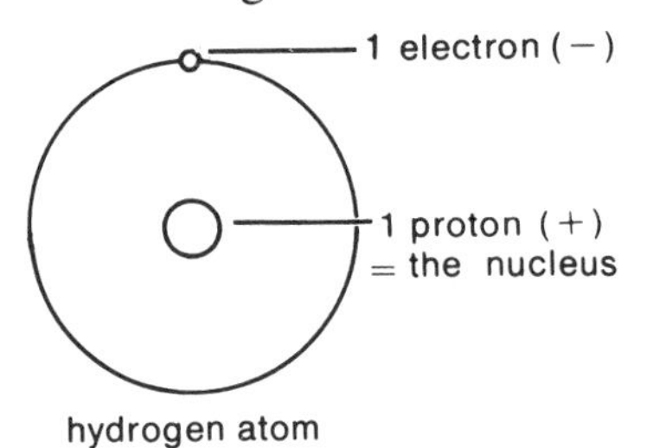

hydrogen atom

The hydrogen atom consists of a negatively charged electron rotating around a positively charged proton. All other elements are expansions and complications of this simple scheme. Hydrogen is the primal stuff of the universe.

Being the simplest element, hydrogen is also the lightest—so light, in fact, that the earth's gravity cannot retain it. When the earth began, hydrogen must have been abundant in the atmosphere. Only traces of it remain now.

"Inflammable air," Cavendish called it. Just how inflammable was demonstrated unforgettably in 1937, when the hydrogen-filled airship *Hindenburg* was incinerated and thirty-six people died at Lakehurst, New Jersey, in less than five minutes.

Free (uncombined) hydrogen is rare on earth, though abundant in the rest of the universe. But combined hydrogen is abundant on earth. Water, after all, is $H_2O$. Cooking gases, petroleum, gasoline, mineral oil, vaseline are all "hydrocarbons," compounds of hydrogen and carbon. Proteins, fats, and carbohydrates, which we eat, which we are, all contain hydrogen.

Hydrogen is what makes acids "acidic"—not oxygen, as Lavoisier thought. In water solution, the hydrogen of acids loses its single electron which leaves just the proton behind. The positively charged proton is called a hydrogen ion. The number of $H^+$ ions in a solution is an exact measure of its acidity.

Boyle, and later Cavendish, prepared hydrogen in their laboratories by treating metals with an acid:

$$\underset{\text{(iron)}}{Fe} + \underset{\text{(sulfuric acid, dilute)}}{H_2SO_4} \rightarrow H_2\uparrow + FeSO_4$$

Not all metals will react with acids in this way. In fact, metals can be ranked according to how reactive they are, with hydrogen as a reference point, in the so-called Activity Series of Metals on page 70.

A surprising amount of information is contained in this list. It explains why one must not throw water on a magnesium flare or incendiary bomb, or even on a burning airplane containing magnesium alloys. Releasing free $H_2$ will not help put out a fire.

The series also shows how readily metals lose electrons: potassium most of all, gold least. From this one can predict how a given metal will behave in the presence of an electrical current, in batteries, and in chemical reactions with other metals. Any metal in solution can be forced out of solution (displaced) by

|  |  |  |
|---|---|---|
| potassium | (K) | VERY REACTIVE: liberate $H_2$ even from cold water: $2K + 2H_2O \rightarrow H_2\uparrow + 2KOH$ |
| calcium | (Ca) | |
| sodium | (Na) | |
| magnesium | (Mg) | REACTIVE: liberate $H_2$ from hot water, steam, and of course acids |
| aluminum | (Al) | |
| zinc | (Zn) | |
| chromium | (Cr) | |
| iron | (Fe) | |
| nickel | (Ni) | |
| tin | (Sn) | SLOWLY REACTIVE: liberate $H_2$ from acids only |
| lead | (Pb) | |
| HYDROGEN | | |
| copper | (Cu) | UNREACTIVE: will not liberate $H_2$ from acids |
| mercury | (Hg) | |
| silver | (Ag) | |
| platinum | (Pt) | |
| gold | (Au) | |

adding one of the metals above it in the activity series. For example, photographic films contain silver salts which are washed into the solutions used for developing. By adding a metal above silver on the list the silver can be salvaged; the added metal will go into solution, "pushing out" the silver in the solid form.

## Saturated and Unsaturated Fats

Fats and oils contain long chains of carbon atoms. Oils pressed from olives, corn, cottonseed, and peanuts contain a number of double ("unsaturated") bonds:

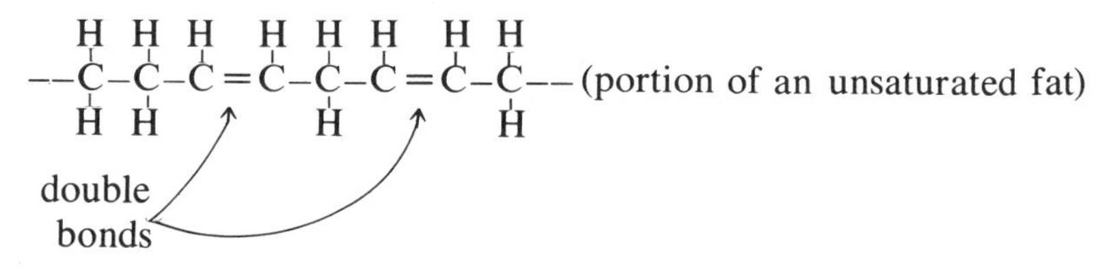

The double bonds are chemically reactive. Sometimes this is useful. The double bonds in linseed oil react with air, forming a tough protective film over oil paintings. But similar reactions in natural oils used for food may produce rancidity.

By treating such oils with hydrogen, the double bonds are "saturated" with additional hydrogen:

```
  H H H H H H H H
  | | | | | | | |
--C-C-C-C-C-C-C-C--    (portion of a saturated fat)
  | | | | | | | |
  H H H H H H H H
```

The original oil has now become a solid white fat which will keep indefinitely without change. "Hydrogenation" (the process of saturating an unsaturated fat or oil with hydrogen) raises the melting point. Thus a manufacturer with gallons of rancid cottonseed oil that no one wants can easily and cheaply transform it into Crisco or margarine, odor free, long lasting.

Medical evidence strongly suggests that saturated fats favor heart disease and arteriosclerosis. But this has not outweighed the enormous convenience hydrogenation represents to the manufacturer. After a year on the grocer's shelf, a "natural" peanut butter may taste odd, but a hydrogenated peanut butter will not —or at least no odder than it did to begin with.

Since federal law does not require an honest statement of the degree of saturation of the product, some manufacturers have taken to advertising margarines "high in polyunsaturates" while offering a product little different from lard except in its higher price.

# 12. Gases, Liquids, and Solids

Alchemy and the phlogiston theory were overthrown by the discoveries of the "gas chemists" of the seventeenth and eighteenth centuries. The next advance in chemistry—atomic theory—also resulted from experiments with gases. What made gases such instructive materials to work with at that time?

In the first place, many of the gases are elements, the simplest form of matter. At the beginning of the nineteenth century, a chemist could learn more from studying how $H_2$ and $O_2$ combine to form water than from trying to make sense out of a material as complex as morphine, $C_{17}H_{19}NO_3 \cdot H_2O$, which was isolated in 1803. Gases were easily and cheaply prepared in the laboratory. They were pure. Their weights and volumes could be measured with simple equipment. Their chemical reactions tended to be rapid and complete.

The physical properties of gases were also being studied.

Boyle had shown that gases are supremely "squeezable"; the more pressure applied to a gas, the less space it takes up.

Boyle, like Newton, believed in atoms. He knew that a liquid or a solid could be squeezed forever without changing its volume in the slightest, so to explain the "springiness" of a gas he claimed that the atoms behaved like invisible coiled springs. When a gas was compressed and its atoms pushed together, the atoms pushed back. "Each corpuscle endeavours to beat off all others from coming within the little sphere requisite to its motion," Boyle wrote in *A Defence of the Doctrine touching the Spring and Weight of the Air* (1662).

The French physicist Jacques Alexandre César Charles (1746–1823) took part in the world's first balloon ascents in 1783. At first the balloon was filled with heated air, later with hydrogen. J. A. C. Charles expressed in mathematical terms the fact that gases expand when heated. This is what makes hot air lighter than ordinary air; hot air has expanded, so that a cubic foot of hot air contains fewer atoms than a cubic foot of ordinary air.

Gases show other temperature effects. When a gas is compressed, it not only takes up less volume, it also gets hot. Inflating a bicycle tire with a hand pump makes both the tire and the pump hot. The atoms are pushed closer together, so that they collide more often, creating heat.

The reverse of this situation occurs when a gas is suddenly released from compression. The expanding gas cools, because the atoms suddenly stop banging into one another. This is why aerosol sprays always feel cold as they emerge from a pressurized can.

These laws are neatly exploited in the electric refrigerator. The motor compresses the refrigerant gas, which heats up but is circulated through a cooling system. When the cooled, compressed refrigerant is allowed to expand again, a drop in temperature occurs.

Rapid mixing and spreading is another characteristic of

gases. A tear-gas cannister can render streets and buildings uninhabitable within minutes. The Scottish chemist Thomas Graham (1805–1869) studied this property and found that the speed with which a gas spreads depends on its weight. As might be expected, a heavy gas does not circulate as fast as a light gas.

Newton had described atoms as repelling one another, but not moving. The rapid diffusion of gases strongly suggested that the atoms most certainly were moving. Daniel Bernoulli (1700–1782), a Swiss mathematician, added the property of constant movement to his view of gas atoms.

The idea that movement had something to do with heat began to emerge. Benjamin Thompson (1753–1814), an American Tory whose services to the Holy Roman Empire led to his being given the title Count Rumford (from the town of Rumford, later Concord, New Hampshire), studied the intense heat that developed when pieces of brass were bored out for cannons. By enclosing the brass in water during the boring, he found, as he reported in *Philosophical Transactions* (1798), that "at 2 hours and 30 minutes it ACTUALLY BOILED!" More important, he concluded that there was no way for heat to be communicated "except it be MOTION."

Count Rumford finally settled in Paris, where he married Lavoisier's widow in 1805. The lady conducted a few heat researches of her own; at the climax of a marital dispute, she poured boiling water on the Count's favorite rose bushes. The two separated after four stormy years.

As the nineteenth century progressed, the idea of atoms in constant chaotic motion evolved into something known as the kinetic theory of matter. "Kinetic" refers simply to motion; it comes from the same Greek root that gave us "cinema."

Moving atoms cause gases to spread. They cause gases to heat up when compressed, due to the increased friction. Since heat is energy, a heated gas expands because its atoms move faster. Another gas law, Graham's Law, expressed mathematically that heavier gas atoms move and diffuse more slowly than light gas atoms; the same amount of energy applied to a heavier atom moves it less.

Cooling is a loss of heat energy; when a gas is cooled, it liquefies. Atoms in a liquid still move and collide, but much less vigorously than in a gas; occasionally a more active atom flies off into space, or evaporates. When even more energy is removed by cooling, the liquid solidifies. The atoms in a solid are very close together and move hardly at all; often they are arranged in regular patterns, or crystals.

The kinetic theory explained the workings of the various Gas Laws (Boyle's Law, Charles's Law, Graham's Law). Later it helped explain how chemicals act in solution. But all this is physics, not chemistry. What took place in a chemical reaction remained as mysterious as ever.

A few firm facts about chemical reactions were known at the beginning of the nineteenth century. One was that elements always react with one another in the same proportions.

To synthesize water from hydrogen and oxygen, eight times more oxygen (by weight) than hydrogen is needed. One ounce of hydrogen and one ounce of oxygen will give one-eighth ounce water, plus seven-eighths ounce unreacted hydrogen. One ounce of hydrogen and eight ounces of oxygen give nine ounces of water, with nothing left over. Two ounces of hydrogen and eight ounces of oxygen give nine ounces of water, plus one ounce of unreacted hydrogen.

"The Law of Definite Proportions" (*Le Principe des Proportions Définies*) was stated in 1799 by the French chemist Joseph Proust (1754–1826) and was widely accepted. Proust left France at the start of the Revolution and settled in Madrid. Violence followed him even there; an anti-French mob sacked his laboratory during the 1808 siege of Madrid by Napoleon's troops.

Napoleon's official chemist, Claude Berthollet (1748–1822), disagreed with Proust's Law. He thought a compound could have many different compositions. Water from the Nile, for example, might have more hydrogen in it than water from the Seine. This disruptive and mind-boggling idea was carefully refuted by Proust. The controversy is a model of civilized disputation, very different from the kind of free-for-all that *"les auteurs anglo-saxons"* often indulged in. With the memory of Lavoisier still green, the French government was pleased to honor Proust and to award him a pension.

Against this background, John Dalton in 1808 published his *New System of Chemical Philosophy,* one of the classic documents of chemistry.

# 13. Atoms and Molecules

*Atoms are blocks of wood, painted various colors, invented by Dr. Dalton.*

——anonymous student, quoted by C. Schorlemmer, *The Rise and Development of Organic Chemistry*

If you cut a piece of gold in half, then cut one of the halves in half, and so on, how long could you keep doing that, assuming you had an ideal knife?

The philosopher Anaxagoras (500?–428 B.C.) thought you could keep dividing the piece of gold forever, and Aristotle supported him. But Democritus (c. 460–c. 370 B.C.) thought that eventually you would reach a particle of gold so tiny that it could not be cut in half. He called this particle *a-tomos* (not cut, not cuttable).

The Democritan atom was rejected by most other Greeks. However, the philosopher Epicurus (342–272 B.C.) made it part of his teaching. And Epicurus's most extravagant admirer, the Roman poet Lucretius (99–55 B.C.), rhapsodized about atoms at great length in *De Rerum Natura* (On the Nature of Things).

During the Middle Ages atomism was in disrepute. Lucretius had been an atomist, and the Church disapproved of Lucretius because, in trying to free man from dependence on the gods, he had written the famous line: *Tantum religio potuit suadere malorum* (Such are the heights of wickedness to which men are driven by religion). In China and India, however, atomism flourished. The great Jewish philosopher Maimonides (1135–1204) described an Islamic theory in which Allah perpetually recreates every atom of the universe from moment to moment.

Although both Newton and Boyle believed in atoms (Boyle called them "corpuscles"), the idea of such particles remained abstract and of little use to chemistry, until the English schoolteacher John Dalton (1766–1844) thought of adding two significant qualities: individuality and weight.

A Quaker, tough-minded, patient, physically strong, Dalton had supported himself from the age of twelve on. For most of his life he was a teacher; meetings of the Manchester Literary and Philosophical Society were his chief pleasures. He avoided marriage as "too distracting."

In 1794 he read to the Manchester Society a paper about color blindness, a condition from which he suffered. He theorized that it was caused by the fluid in the eyeballs being tinted. However, after his death one of his eyes was removed and opened, and the fluid proved to be colorless. The eyeball is still preserved at the University of Manchester.

By 1803 he had formed the basic ideas of his atomic theory. He presented them to the Manchester Society; in 1808 he published them in *A New System of Chemical Philosophy*.

Just how Dalton arrived at these ideas is still uncertain—he gave several contradictory accounts. He seems to have done most of his experimentation after announcing the theory, rather than before. This is deduction: forming a general rule and then applying it to specific cases. Science by tradition is inductive: it collects evidence from specific cases and then forms a general rule. (Crime detection is also inductive, but because Sir Arthur Conan Doyle in 1889 began a Sherlock Holmes story, *The Sign of Four,* with a chapter entitled "The Science of Deduction," much of the English-reading public has believed ever since that detectives spend all their time "deducing.")

In actual practice the process of scientific discovery is much more complicated. Scientists do follow "recipes," but they also do anything else they think may possibly work. Chance and opportunism are involved, also imagination, the unconscious, and intuition.

Dalton's main asset was intuition, rather than a mastery of formal logic. According to Leonard Nash (*Harvard Case Histories in Experimental Science*), Dalton had "a genius for forging ahead by *guessing* right when insufficient criteria were available to support a completely reasoned judgment."

Dalton's atomic theory proposed that:

(1) all matter is made up of atoms;
(2) the atoms of any single element all have the same shape, size, weight, and behavior;
(3) the atoms of one element differ from the atoms of all other elements in shape, size, weight, and behavior;
(4) atoms are never created or destroyed; in chemical reactions, they combine, separate, or regroup;
(5) when two elements combine to form a compound, they do so in simple ratios, 1:1, 1:2, 1:3, and so on (since only whole atoms combine, not fractions of atoms).

These ideas may not seem electrifying today, but they set chemical thinking in the nineteenth century firmly on the right track. There were no more attacks on the Law of Definite Proportions.

Most accounts of Dalton stop tactfully at this point. However, Dalton lived on for another forty years, during which time his views became permanently fixed. The tenacity that had helped him during his early years now made him cling to errors in his theory, even when the information to correct them became available.

Part of his problem was semantic. The term "atom" is used today only for a single uncombined atom of an element. The moment an atom combines with anything—even another atom like itself—the particle that results is called a "molecule":

$$H = \text{a hydrogen atom}$$
$$H_2 = \text{a hydrogen molecule}$$
$$H_2O = \text{a water molecule}$$

The term molecule had long been in use to describe any small "body," and in 1811, it was precisely defined by the Italian physicist Amedeo Avogadro (1776–1856), who showed how its use would avoid confusion. But Dalton would have none of it. Having invented an atomic theory, he applied the term "atom" to everything. Thus he talks about "the atoms of compounds," "water atoms," how the "atoms" are arranged in "atoms of alcohol," and so on, blurring the distinction between elements and compounds in the reader's mind, and occasionally in his own.

One great feature of the atomic theory was the ascribing of weight to elemental atoms. Dalton knew he could not get the actual weights of different atoms, but he could get the relative weights; he could determine how much heavier or lighter an element was when compared to another element. He took the lightest element, hydrogen, as his standard and gave it an atomic weight of 1. Then his troubles started.

He assumed the reaction of hydrogen and oxygen to form water was

$$H + O \rightarrow HO$$

but actually two atoms of hydrogen combine with each oxygen atom. When he found that 1 ounce of hydrogen reacted with 8 ounces of oxygen, he called the atomic weight of oxygen 8; but since the 1 ounce represents two hydrogen atoms, a single hydrogen atom is just ½ ounce, and the atomic weight of oxygen is actually 16.

Moreover, hydrogen and oxygen occur as biatomic molecules, so the correct reaction is

$$2H_2 + O_2 \rightarrow 2H_2O$$

Even when Dalton hit on the right formula for a compound, he still went astray because of the biatomic molecules. The formula for nitric oxide is NO, as he assumed, but the reaction is not

$$N + O \rightarrow NO$$

The correct reaction is

$$N_2 + O_2 \rightarrow 2NO$$

At first, Dalton had no way of knowing these facts. But as other scientists gave him evidence, Dalton preferred to throw out the evidence rather than change his original ideas.

## Dalton and Gay-Lussac

Joseph Louis Gay-Lussac (1778–1850) was a distinguished chemist, professor, peer of France, and balloon enthusiast. (In 1804 he ascended, alone, more than four miles above Paris in a hydrogen balloon.) Both he and Dalton published experiments on the expansion of gases with heat (1801–1802) and were rivals for the honor of having this particular gas law named after them. However, J. A. C. Charles had made the discovery earlier, though without publishing it, so the law was named after him.

Guy-Lussac, having found a young lady reading a chemistry book in a drygoods store, married her in 1808 (the year of Dalton's *New System*). He then settled down and made a serious study of the chemical reactions of gases. He measured not just the weights of the gases, but their volumes (the space they occupied) as well.

The volumes of the reacting gases were always in simple ratios, 1:1, 1:2, . . . , and this certainly supported "Dalton's ingenious idea." Hydrogen chloride, for example, was formed from one pint of hydrogen and one pint of chlorine.

Some of the results were puzzling, however. One pint of nitrogen reacted with one pint of oxygen. If a pint of nitrogen contains 1 trillion N atoms and a pint of oxygen contains 1 trillion O atoms, one would expect them to combine and form a pint of 1 trillion NO molecules. But two pints were obtained.

Still more puzzling was the formation of water. Gay-Lussac found that two pints of hydrogen combined with one pint of oxygen (as would be expected from the formula $H_2O$), but they produced two pints of water vapor. This seemed to imply that the oxygen atom was being split. And as Dalton used to say, "Thou knows no man can split the atom."

The explanation, of course, was the biatomic molecule. The pint of oxygen contained 1 trillion $O_2$ *molecules* and these actually *did* split to provide 2 trillion O atoms for 2 trillion water molecules. Here is what actually happened in the Gay-Lussac reactions:

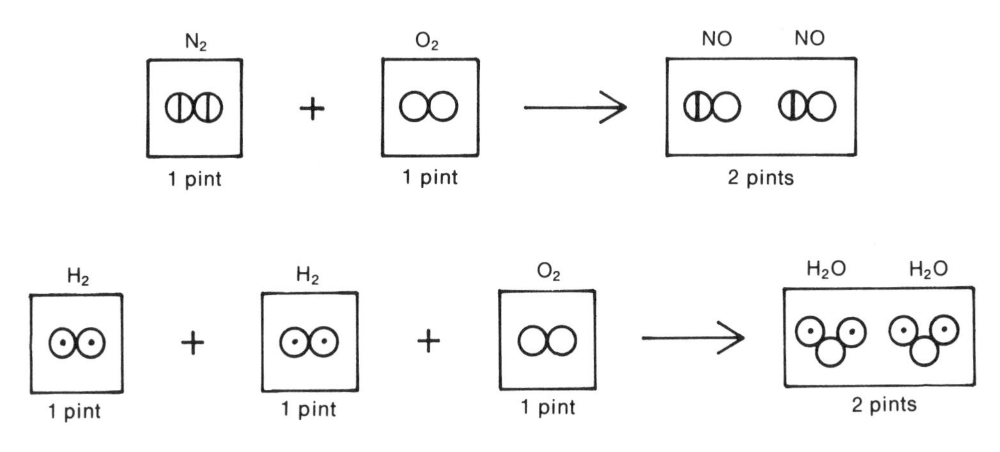


Dalton attacked Gay-Lussac's work in 1810 and remained opposed to it all his life. He went so far as to say that Gay-Lussac's Law of Combining Volumes depended on "the inaccuracy of our experiments," by which he meant the inaccuracy of Gay-Lussac's experiments.

## Dalton and Avogadro

A year later, Avogadro wrote a famous article in the *Journal de Physique*. In it he praised both Dalton and Gay-Lussac and showed how their work could be reconciled. Avogadro proposed that:

(1) gaseous atoms can link up to form biatomic ($O_2$) or triatomic ($O_3$) molecules, each of which takes up the same space as a single atom. This opened the way to the correct explanation for Gay-Lussac's experimental results.

(2) equal volumes of different gases contain the same numbers of atoms (or molecules), when the temperature and pressure of each are the same. This shows a simple way of determining the atomic weights of gaseous elements: simply weigh equal volumes of hydrogen (or whatever gas is used as a standard) and the test gas. Whatever weight is found for hydrogen, the weight for oxygen will always be 16 times greater. The atomic weight of oxygen is therefore 16.

Dalton was particularly opposed to the second point. He was convinced that the atoms of different elements occupied different amounts of space; therefore equal volumes of different gases would have different numbers of atoms.

As for the first point, Dalton retained Newton's idea that gas atoms repelled one another. If they did, how could they get close enough to link up as molecules?

And so the reasonable, unifying, and enormously useful Avogadro hypothesis was neglected for nearly fifty years. Not until the Italian chemist Stanislao Cannizzaro (1826–1910) vigorously promoted it at an international chemical congress at Karlsruhe, Germany, in 1860 was it at last accepted. For some, the revelation had almost religious overtones. The German chemist Lothar Meyer (1830–1895) wrote: "The scales fell from my eyes, doubts vanished, and a feeling of calm certainty came in their place."

One of the consequences of Avogadro's rehabilitation was announced stop-press fashion in *Scientific American* in 1861:

"The formula for water, according to recent discoveries, must soon be changed from HO to $H_2O$."

## Dalton and Berzelius

The Swedish chemist Jons Jakob Berzelius (1779–1848) had an extraordinarily productive and influential career. At seventeen, without parents or money, he left school with, according to J. R. Partington in *A History of Chemistry,* "the usual report of headmasters on pupils of genius that there was 'little hope for him.' " Working in a small kitchen laboratory, Berzelius managed to discover several elements, to invent the system of chemical symbols that we use today, to make basic discoveries in electrochemistry, to start a chemical journal, to correspond with all the major chemists of the day, and to train so many pupils so successfully that from then on it is hard to find any chemist of note who did not study with Berzelius or at least study with someone who studied with Berzelius.

The aging poet Goethe was one of Berzelius's friends. Not surprisingly, the author of *Faust* rejected Dalton's atomic theory. He preferred to continue trying to make gold in his small blast furnace. Berzelius, on the other hand, welcomed the theory. By 1830 he had prepared a table of atomic weights with values nearly as accurate as those used today. As tactfully as he could, he tried to induce Dalton to see the value of Gay-Lussac's researches.

Dalton had proposed a system of chemical symbols, realizing that alchemical symbols were no longer suitable. For simple compounds, Dalton's symbols were helpful. Carbon dioxide was

But alum (potassium aluminum sulfate) looked like this:

Berzelius devised a simpler, less arbitrary system of symbols, which is the basis of that now used. Dalton found these symbols "horrifying," resembling Hebrew. A heated discussion of the Berzelius system preceded the first stroke suffered by Dalton in his seventies.

A few years earlier, Berzelius at fifty-six had been on the verge of collapse from overwork and depression after the death of his wife. Marriage to a twenty-four-year-old girl produced a remarkable cure, however, and he had a dozen more years of active professional life.

# 14. Water

*The constancy of the internal environment* (le milieu intérieur) *is the necessary condition of the free life.*
——Claude Bernard, *Leçons sur la Chaleur Animale*

Life arose in the sea, and the only way an animal could leave it was by carrying his own small private ocean around with him. Animals evolved a system of circulating fluids within the body—blood and lymph—which have much in common with sea water.

Something similar happened to plants when they invaded land areas. But they remained dependent on fluids in the soil, a kind of underground sea.

Human beings are about two-thirds water. The first table on the next page shows how they rank with other biological materials.

Since water is an excellent solvent, any water found in nature will have things dissolved in it. Even freshly collected rain

| *Material* | *Percent water (by weight)* |
|---|---|
| cucumbers | 96 |
| watermelons | 92 |
| milk | 87 |
| apples | 84 |
| potatoes | 78 |
| steak | 74 |
| humans | 66 |
| cheese | 40 |
| bread | 35 |

water will contain pollen, dust, dissolved gases. Oceans contain about 3.5 percent dissolved solids (mostly NaCl). The Great Salt Lake in Utah contains 22 percent dissolved solids (again mostly NaCl). When an ocean dries up, it leaves a large deposit of salt behind. Parts of the salt bed in Death Valley are 5,000 feet thick.

The main reason that oceans are saltier than fresh water is that they have been around longer; being older, they have had more time to dissolve the materials they contact.

Water is harder to heat or to cool than are many other materials. This quality is expressed in exact terms as "specific heat" —the calories required to raise the temperature of 1 gram ($1/28$ ounce) of a substance 1° C.

| *Material* | *Specific heat* |
|---|---|
| iron | 0.10 |
| glass | 0.12 |
| aluminum | 0.21 |
| water | 1.00 |

An empty aluminum saucepan placed on a stove burner heats almost instantly; water takes several minutes. This sluggishness of water in changing temperature has far-reaching climatic effects.

Cities on a coast tend to have even temperatures all year around. San Francisco, "the city of perpetual autumn," has mild winters and cool summers. There are some seasonal changes, of course, but nothing like those found in Sacramento (100 miles northeast) or Fresno (200 miles southeast).

The stabilizing effect of large bodies of water results from the temperature lag. By the end of summer, coastal waters have become warm and they warm the cold winter air as it passes over them. By the time the coastal waters finally cool down, summer returns with warm air which the ocean now cools.

Hawaii is entirely surrounded by a vast expanse of ocean with a temperature nearly always 80° F.; the average year-round air temperature of Hawaii is the same.

Water has another curious physical property. Like most materials, water "shrinks" and becomes more dense as it cools. But when the temperature falls below 4°, water reverses itself and expands, becoming less dense. Since water becomes lighter just before it freezes, only the tops of lakes freeze in winter. Because of this unique property and because ice is a good insulator, water at the bottom of lakes remains slightly warmer and unfrozen, allowing many forms of life to survive the winter. A less happy consequence of this slight expansion of water before freezing is seen in burst water pipes and car radiators.

Water is a convenient standard. The standard pound is based on the weight of 7,000 plump grains of wheat and kept locked up in Washington, D.C. But when water is used as a standard, there is no need to make trips to Washington. The weight unit of the metric system, the kilogram, is taken to be the weight of 1 liter of water at 4° C., when it is heaviest. Similarly in the centigrade (hundred-step) temperature scale, the freezing point of water is taken as 0° and its boiling point as 100°.

## Solutions

The only way the various chemical compounds in living organisms can move about and reach one another is by being dissolved in water. Since nearly all biochemical reactions take place in water solution, it is worth learning how materials behave when they dissolve in water.

So important an area of human knowledge must have a special terminology. Instead of saying that gasoline and water do not mix, one must say that they are *immiscible*. Alcohol and water can be mixed, needless to say, so they are *miscible*. When a lump of sugar is dissolved in a cup of hot coffee, the sugar is called the *solute* (the thing being dissolved) and the coffee is called the *solvent* (the thing that does the dissolving).

Which is the solvent and which the solute in a combination of 50 percent alcohol and 50 percent water? Actually, it is optional. You can decide on the basis of which compound you are more interested in, or you can toss a coin.

A *solution* is any homogenous mixture of two or more kinds of atoms, molecules, or ions. The atmosphere is a solution of gases in gases. An alloy is a solution of metals in metals. Here we are concerned only with aqueous (water) solutions:

a solid (sugar, salt, protein) in water
a liquid (alcohol, glycerine) in water
a gas (carbon dioxide) in water

Champagne represents all three types of aqueous solution, since it contains sugar, alcohol, and $CO_2$ evenly mixed in water.

Solutions have special properties. They will not "settle out," like muddy water. They can be filtered through the finest filter without being changed (that is, the coffee is just as sweet as before). The solute is evenly distributed in the solvent (a spoonful of coffee from the bottom of the cup is just as sweet as a spoonful from the top).

Temperature affects the amount of a substance that will dissolve. As a rule, the hotter the water, the greater the amount of a given solid will dissolve in it and the lesser the amount of a gas will dissolve.

When adding solute, a point is eventually reached when no more of the solute will dissolve in the solvent. The solution is now *saturated*. A crystal of salt added to a saturated salt solution will remain a crystal. If the temperature of the solution is raised, a new solubility level is established and the crystal will dissolve.

## Molarity and Normality

Percent by weight is a common way of showing exactly how strong a solution is. Two ounces of boric acid in 100 ounces of water is a 2 percent solution of boric acid.

Chemists prefer to use a system that also shows at a glance how strong a solution will be in a chemical reaction. Weight is no help here; the number of molecules is the information needed—not so much an actual count of molecules in billions and trillions, but a way of comparing the numbers of molecules in different solutions.

First of all, the idea of atomic weights must be extended to molecular weights. In chapter 13, atomic weights were described as relative weights, not the actual weight of atoms:

| *Element* | *Atomic weight* |
|---|---|
| H | 1 |
| O | 16 |
| N | 14 |
| C | 12 |

This simply means that, since equal volumes of gas contain the same numbers of atoms or molecules (Avogadro's hypothesis), if a pint of oxygen weighs 16 times as much as a pint of hydrogen, then a single oxygen atom weighs 16 times as much as a single hydrogen atom. This works for compounds as well as for elements. Adding together the atomic weights in a compound gives the molecular weight:

| | | | |
|---|---|---|---|
| Ammonia ($NH_3$) | = | N | 14 (atomic weight) |
| | | H | 1 (atomic weight) |
| | | H | 1 (atomic weight) |
| | | H | 1 (atomic weight) |
| | | | 17 (molecular weight) |

| | | | |
|---|---|---|---|
| Carbon dioxide ($CO_2$) | = | C | 12 (atomic weight) |
| | | O | 16 (atomic weight) |
| | | O | 16 (atomic weight) |
| | | | 44 (molecular weight) |

When actually tested, a pint of $NH_3$ weighs 17 times as much as the standard hydrogen pint, while a pint of $CO_2$ weighs 44 times as much. As long as the atomic weights are accurate, the molecular weights calculated from them are also accurate.

A solution consisting of the molecular weight of a compound *in grams* dissolved in 1 liter of water is called a *one-molar solution* (1 M). A 1 M solution of $NH_3$ is 17 grams of $NH_3$ in a liter of water. If the molecular weight of salt (NaCl) is 58,

| | | |
|---|---|---|
| 1 M NaCl | = | 58 grams NaCl per liter |
| 2 M NaCl | = | 116 g NaCl per liter |
| 10 M NaCl | = | 580 g NaCl per liter |
| 0.1 M NaCl | = | 5.8 g NaCl per liter |

Equal amounts of solutions of the same molarity contain the same numbers of molecules. A drop from a 1 M NaCl solution contains the same number of NaCl molecules as a drop from a 1 M lysergic acid diethylamide solution contains LSD molecules, or a 1 M potassium cyanide solution KCN molecules.

In reactions between one molecule of one compound and one molecule of another compound, equal amounts of the same molarity solutions of the two compounds will react completely with each other. If one molecule of one compound reacts with three molecules of the second compound, three times as much solution of the second compound with the same molarity can be used or the same amount of a solution with a molarity three times as strong.

Reactions between acids and bases are a special case. The concentration of the H and OH parts of the molecule is more important there than is the molecule as a whole. So "normal" (N) solutions instead of molar solutions are used:

$$1\ M\ HCl = 1\ N\ HCl$$
$$1\ M\ H_2SO_4 = 2\ N\ H_2SO_4$$
$$1\ M\ H_3PO_4 = 3\ N\ H_3PO_4$$

Thus equal amounts of solutions of the same normality always contain the same amounts of H and OH (even for those compounds with two and three times more H or OH built into the molecule).

The term "weight" as used in this discussion is not strictly correct. Weight is tied to gravity. A dumbbell weighing 25 pounds on the earth's surface may not weigh anything in a spaceship. Yet the dumbbell has not changed as far as the amount of matter present is concerned. The term "mass" is used as a measure of the quantity of matter present, independent of gravity. The gram is the basic unit of mass.

## Ionization

Pure distilled water does not conduct electricity well. If you were swimming in an unpolluted alpine lake and lightning struck it, you would probably escape electrocution. Similarly, a lamp will not light if the electrical circuit is broken and the two broken ends are put in a glass of pure water. But if a pinch of salt is dissolved in the water, the lamp will light.

Salts have another odd property in solution. When a compound is added to water, the freezing point of the water is lowered. This is the basis of anti-freeze used in car radiators. Salts are too corrosive to use as anti-freeze materials, but most of them are twice as effective as the materials which are used (alcohols, glycols). Molar solutions are presumed to contain the same numbers of molecules. Yet a 1 M alcohol solution freezes at $-2°$ C., while a 1 M salt solution freezes at $-4°$ C.

In 1884 the Swedish chemist Svante Arrhenius (1859–1927) tried to explain this phenomenon in his Ph.D. thesis. He suggested that each salt molecule split apart in water into two electrically charged particles:

$$NaCl \rightarrow Na^+ + Cl^-$$

He called the charged particles "ions," from the Greek verb meaning "to go." Thus an ion is a "go-er," or more poetically a wanderer. Ion, a son of Apollo, and the hero of Euripedes's play of that name, was the legendary ancestor of the Ionian Greeks.

Arrhenius's ionization theory failed to impress his professors. They gave him the lowest possible grade short of failure. But twenty years later he received the Nobel Prize for his theory.

Alcohols do not ionize appreciably. Most salts, however, ionize 100 percent. In the case of NaCl, each ion is effective in lowering the freezing point, which is why salt is twice as effective as alcohol for this purpose.

An electrical current does not readily cross an electrically neutral space of water. But if the water is filled with mobile, electrically charged ions, the current crosses easily.

The ionization theory explains why acids differ in strength even when their normal solutions are the same. Acid strength depends on numbers of $H^+$ ions formed. A 0.1 M solution of hydrochloric acid (HCl) is 100 percent ionized, so the $H^+$ concentration in this solution is also 0.1 M:

$$\begin{array}{ccccc} HCl & \rightarrow & H^+ & + & Cl^- \\ 0\% & & 100\% & & 100\% \end{array}$$

But a 0.1 M solution of acetic acid only ionizes 1.3 percent, so the $H^+$ concentration in this solution is nearly 100 times less, or 0.001 M:

$$\begin{array}{ccccc} CH_3COOH & \rightarrow & H^+ & + & CH_3COO^- \\ 98.7\% & & 1.3\% & & 1.3\% \end{array}$$

## Electrochemistry

Substances which ionize and thus conduct electricity are called *electrolytes*. The "lyte" part comes from the Greek *lysis,* destruction. (Aristophanes's Lysistrata was the "destroyer of battles.") When hair is removed by electrolysis, the hair follicle is destroyed, presumably. The electrolysis of water destroys it in the sense of breaking it down into hydrogen and oxygen. An electrolyte is "destroyed" by electricity, or at least pulled apart into ions.

Electrical energy can be converted into chemical energy, and vice versa, with interesting results in both cases.

For convenience, an electrical current may be defined as a flow of electrically charged particles. If the particles are electrons, as in a copper wire, there is metallic conduction. If the particles are ions, there is electrolytic conduction.

In a battery, chemical energy is turned into electrical energy by means of two simultaneous reactions. In one cell zinc metal goes into solution, releasing electrons; in the other cell copper ions $Cu^{++}$ come out of solution as metallic Cu, taking up electrons. The flow of electrons from the zinc cell to the copper cell has a driving force of about 1.1 volt.

$$\begin{array}{ll} Zn \rightarrow Zn^{++} + 2e & \text{(half-reaction \#1)} \\ 2e + Cu^{++} \rightarrow Cu & \text{(half-reaction \#2)} \\ \hline Zn + Cu^{++} \rightarrow Zn^{++} + Cu & \text{(total reaction)} \end{array}$$

When the situation is reversed and electrical energy is put into the system, chemical work is done. For example, sodium is now manufactured commercially by the electrolysis of molten sodium chloride; chlorine gas is released at one electrode, while sodium is deposited at the other:

$$2Na^+ + 2Cl^- \rightarrow 2Na + Cl_2\uparrow$$

Reactions like this are the basis of the electroplating industry; they permit thin layers of expensive or durable metals to be applied to cheaper or less lasting metals. Jewelry and tableware can be silver-plated by hanging the article from the negative electrode in a solution of silver nitrate ($AgNO_3$). As the electrical current flows, silver ions in solution plate out as metallic silver:

$$Ag^+ + e \rightarrow Ag$$

To prevent rusting and corrosion, steel cans are electroplated with tin (in $SnCl_2$ solution), and automobile trim is electroplated with chromium (in $CrCl_3$ solution).

## Acids and Bases

"Acid" comes from the Latin word for "sour," which is not much help in defining an acid chemically. Lavoisier's theory that oxygen makes acids "acidic" was proved wrong by the discovery of hydrochloric acid (HCl). Chemists then saw that all acids contained hydrogen. But many other compounds contain hydrogen without being in any way acidic: natural gas ($CH_4$), ammonia ($NH_3$), sugar ($C_{12}H_{22}O_{11}$), and water itself ($H_2O$).

Arrhenius clarified the situation by defining an acid as a compound which gives hydrogen ions ($H^+$) in solution. The four compounds mentioned in the preceding paragraph do not do this to any significant extent, but the familiar acids do:

| | |
|---|---|
| hydrochloric acid | $HCl \rightarrow H^+ + Cl^-$ |
| nitric acid | $HNO_3 \rightarrow H^+ + NO_3^-$ |
| sulfuric acid | $H_2SO_4 \rightarrow H^+ + HSO_4^-$ |
| | $HSO_4^- \rightarrow H^+ + SO_4^{=}$ * |
| carbonic acid | $H_2CO_3 \rightarrow H^+ + HCO_3^-$ |
| | $HCO_3^- \rightarrow H^+ + CO_3^{=}$ |
| acetic acid | $CH_3COOH \rightarrow H^+ + CH_3COO^-$ |

Having two hydrogen atoms in the molecule, sulfuric and carbonic acids ionize in two steps. In the first step they produce

* This symbol represents two minus signs; $SO_4^{--}$ and $SO_4^{2-}$ are alternative usages.

the bisulfate ($HSO_4^-$) ions and the bicarbonate ($HCO_3^-$) ions respectively (or more correctly the hydrogen sulfate and hydrogen carbonate ions). In the second step they form the sulfate ($SO_4^=$) and carbonate ($CO_3^=$) ions respectively. Carbonic acid and acetic acid are only slightly ionized, so they are classed as weak acids. The negative ions formed by the remaining acids are chloride ($Cl^-$), nitrate ($NO_3^-$), and acetate ($CH_3COO^-$).

Before Arrhenius, bases (alkalis) were even less well understood than acids but they all had one property in common. They could neutralize acids and make even the strongest acid as harmless as water. In fact, bases do just that: they change acids to water. A base is a compound that gives hydroxide ions ($OH^-$) in solution:

sodium hydroxide $NaOH \rightarrow Na^+ + OH^-$
potassium hydroxide $KOH \rightarrow K^+ + OH^-$
ammonium hydroxide $NH_4OH \rightarrow NH_4^+ + OH^-$

Ammonia gas is basic in water because it forms ammonium hydroxide:

$$NH_3 + HOH \rightarrow NH_4OH \rightarrow NH_4^+ + OH^-$$

Sodium carbonate is also basic because of a reaction with water:

$$Na_2CO_3 \rightarrow 2Na^+ + CO_3^=$$
$$CO_3^= + HOH \rightarrow HCO_3^- + OH^-$$

In the neutralization reaction, the hydroxide ion ($OH^-$) of a base combines with the hydrogen ion ($H^+$) of an acid to form water:

$$H^+ + OH^- \rightarrow HOH$$

Water is ionized only to a very slight extent (making it a poor conductor in the absence of electrolytes). In pure water there are 550,000,000 un-ionized HOH molecules for each $H^+$ and $OH^-$ present. When large numbers of additional $H^+$ and $OH^-$ ions are added in a neutralization reaction, most of them snap together into HOH molecules to maintain the proportion of un-ionized to ionized particles found in pure water. Only when there are more $H^+$ than $OH^-$ ions, or more $OH^-$ than $H^+$ ions is the water acidic, or basic, respectively.

A crude way of telling if a solution is acidic or basic is by dipping a piece of litmus paper into it. Litmus is a dye extracted from lichens which turns pink in acid and blue in alkali. A far more exact way of measuring acidity and alkalinity is by means of a pH meter.

A 1 M $H^+$ solution contains 1 gram $H^+$ per liter of water. Pure water is a 0.0000001 M $H^+$ solution (containing 0.0000001 gram $H^+$ per liter). Instead of writing all those zeros, this value can be expressed as $1 \times 10^{-7}$, which means "1 with the decimal point moved 7 places to the left." An even simpler way of expressing this is used in the pH scale; pH is the negative logarithm of the $H^+$ concentration. As the $H^+$ concentration goes up (acidity increases), the pH goes down. The pH of pure (neutral) water is 7.

Here is the entire pH scale:

| *pH* | | *$H^+$ concentration (molarity)* | *$OH^-$ concentration (molarity)* |
|---|---|---|---|
| 0 | strongly ACIDIC | $1 \times 10^{0}$ | $1 \times 10^{-14}$ |
| 1 | | $1 \times 10^{-1}$ | $1 \times 10^{-13}$ |
| 2 | | $1 \times 10^{-2}$ | $1 \times 10^{-12}$ |
| 3 | | $1 \times 10^{-3}$ | $1 \times 10^{-11}$ |
| 4 | | $1 \times 10^{-4}$ | $1 \times 10^{-10}$ |
| 5 | | $1 \times 10^{-5}$ | $1 \times 10^{-9}$ |
| 6 | weakly ACIDIC | $1 \times 10^{-6}$ | $1 \times 10^{-8}$ |
| 7 | NEUTRAL | $1 \times 10^{-7}$ | $1 \times 10^{-7}$ |
| 8 | weakly BASIC | $1 \times 10^{-8}$ | $1 \times 10^{-6}$ |
| 9 | | $1 \times 10^{-9}$ | $1 \times 10^{-5}$ |
| 10 | | $1 \times 10^{-10}$ | $1 \times 10^{-4}$ |
| 11 | | $1 \times 10^{-11}$ | $1 \times 10^{-3}$ |
| 12 | | $1 \times 10^{-12}$ | $1 \times 10^{-2}$ |
| 13 | | $1 \times 10^{-13}$ | $1 \times 10^{-1}$ |
| 14 | strongly BASIC | $1 \times 10^{-14}$ | $1 \times 10^{0}$ |

Keep in mind that each number on the pH scale represents a tenfold change in acidity: that is, pH 5 is ten times more acidic than pH 6, pH 4 is 100 times more acidic than pH 6, and so forth.

Here are the positions of a number of familiar materials on the pH scale:

| *pH* | |
|---|---|
| 0 | hydrochloric acid (1 molar) |
| 1 | |
| 2 | lemons, vinegar |
| 3 | oranges, raspberries, apples |
| 4 | grapes, tomatoes |
| 5 | cabbage, boric acid |
| 6 | corn, potatoes, peas |
| 7 | blood, pure water |
| 8 | eggs |
| 9 | baking soda |
| 10 | milk of magnesia |
| 11 | ammonia |
| 12 | sodium carbonate |
| 13 | |
| 14 | sodium hydroxide (1 molar) |

If one has an unknown sample (the waste being discharged from a paper mill into a river, for example) and wants to know how acidic it is, one slowly adds a base solution whose strength is exactly known until neutrality (pH 7) is achieved. The amount of base added corresponds to the amount of acid in the sample.

If a pH meter is not available for determining the neutral point, an "indicator" dye can be used instead. Many organic dyes change color with changes in pH. Methyl Red, for example, is pink in acid solution but turns yellow at a point close to neutrality.

Mixtures of these dyes are used to prepare various "universal pH papers." When dipped into unknown solutions, the paper turns different colors according to the pH.

If you should spill or swallow a strong acid (or base), do not apply a strong base (or acid). Since no bell rings when neutrality is reached, you may easily overshoot it and do even more damage with your antidote.

Dilution is the safest remedy. Drink water, or pour water on the affected area. Most strong acids and bases are not poisonous when diluted. The following weak antidotes are also safe:

| *For Acids* | *For Bases* |
|---|---|
| INTERNAL | |
| milk of magnesia | vinegar |
| sodium bicarbonate solution | citrus juice |
| milk (the protein = weak base) | dilute acetic acid |
| EXTERNAL | |
| sodium bicarbonate (solid) | vinegar |
| ammonia solution | dilute acetic acid |

## Dehydration

Drying food is one of the oldest ways of preserving it. The bacteria and molds that spoil food need moisture and have trouble growing on dried meat, fish, apricots, figs, and prunes.

Since water is the main component of many foods, dehydration gives a lighter, less bulky product, suitable for carrying on camping trips.

Some materials contain no water at all, however. It is quite impossible to dehydrate alcohol. Nothing is present in 100 percent alcohol except alcohol molecules; if the molecules are changed in any way, the material stops being alcohol. The idea of dissolving a little packet of powder in water and obtaining instant gin is attractive, but alcohol cannot be dehydrated any more than gasoline or gravel can.

Students of semantics and truth in advertising find the saga of Instant Milk instructive. When first introduced, this product was hard to dissolve in water; after furious stirring, one produced a kind of lumpy library paste. Then came the spray-dry process which produced large, easy-to-dissolve crumbs. This put the manufacturer (Borden's) in a serious semantic bind, since the original product had already been called "instant." The problem was solved by calling the new product "faster than instant."

Some chemicals have the property of absorbing water from the air (they are "hygroscopic" or "deliquescent"). Unless stored in a dry atmosphere, a bottle of deliquescent crystals may change into a gooey syrup overnight.

One way of providing a dry atmosphere is the Desiccator, an airtight glass vessel from which the air can be withdrawn by vacuum pump and which contains anhydrous (no water) $CaCl_2$.

Hair absorbs varying amounts of water, depending on the humidity. This is why shampooing and changes in the weather affect the behavior of human hair. Hair is sensitive enough in this respect to be used in instruments for measuring humidity.

## Water and Life

In the living organism, molecules are torn down for energy, synthesized for growth, excreted, absorbed, duplicated, exchanged, all at once, requiring the most exact coordination and split-second timing. Only in water solution can the molecules reach one another rapidly enough. Water is the carrier, the transporter, of these molecules.

Deprived of water, this transportation system breaks down. The blood thickens. Oxygen and nutrients reach the cells less efficiently, waste products accumulate, and death from multiple causes follows: asphyxiation, self-poisoning, starvation.

The operation of taste and smell requires moist surfaces; so does gaseous exchange in the lungs. Water acts as a lubricant for the internal organs, bones, tendons, and joints. The property of water that causes oceans to resist sudden temperature changes does the same thing for living organisms.

A form of life not based on water is hard to imagine. The reactivity of a cell filled with a mixture of dry solids would be virtually zero.

Mixtures of gases might work better. At least the molecules could move around. The organizational problems would be formidable, but they were also formidable for water-based organisms and were solved successfully. The possibility of springing leaks—a problem in water-based organisms—would be far more hazardous in a gas-based organism, not to mention the possibility of exploding.

# 15. Chemical Change

*Green leaves of summer*
*Turn red in the fall,*
*To brown and to yellow they fade;*
*And then they have to die,*
*Trapped within the circle time-parade*
*Of changes . . .*
——Phil Ochs, *Changes*

Buddhists have always recognized change, impermanence, as one of the three basic qualities of life. For 5,000 years the Chinese *Book of Changes* (*I Ching*) has attracted readers, partly because of the skill with which it reflects the flux of the natural world.

The earth turns and the seasons change. Time passes, plants grow and decay, and everything turns into something else.

In chemistry change is shown, very clearly and exactly, by equations. A surprising amount of information is packed into an ordinary chemical equation. For one thing, all the equations in this book "balance." There are the same number of atoms on the left-hand side as on the right-hand side. In chemical reactions, matter is not created or destroyed but merely rearranged.

From a simple chemical equation, one can calculate the actual amounts—in pounds, grams, or tons—that are involved. For example, when the disagreeable gas sulfur trioxide ($SO_3$) reacts with water, sulfuric acid ($H_2SO_4$) is produced:

$$SO_3 + H_2O \rightarrow H_2SO_4$$

The equation balances. There are 2 H atoms, 1 S atom, and 4 O atoms on each side. By consulting a table of atomic weights, one can obtain the molecular weights of the three compounds involved:

*Atomic weights*

H = 1
O = 16
S = 32

*Molecular weights*

| $SO_3$ | | $H_2O$ | | $H_2SO_4$ | |
|---|---|---|---|---|---|
| = | S: 32 | = | H: 1 | = | H: 1 |
| | O: 16 | | H: 1 | | H: 1 |
| | O: 16 | | O: 16 | | S: 32 |
| | O: 16 | | 18 | | O: 16 |
| | 80 | | | | O: 16 |
| | | | | | O: 16 |
| | | | | | O: 16 |
| | | | | | 98 |

Repeating the original equation with the molecular weights added gives:

$$\underset{(80)}{SO_3} + \underset{(18)}{H_2O} \rightarrow \underset{(98)}{H_2SO_4}$$

This shows the proportions (by weight) in which this reaction takes place. One can calculate:

(1) how much $SO_3$ and $H_2O$ are needed to produce any desired amount of $H_2SO_4$.

(If 1,000 tons of $H_2SO_4$ are wanted,

$$\frac{1{,}000}{98} = \frac{x}{80} = \frac{y}{18}$$

$$x = 816 \text{ tons } SO_3$$
$$y = 184 \text{ tons } H_2O)$$

(2) how much $H_2SO_4$ can be obtained from any given amount of $H_2O$ or $SO_3$.
(If, because of a drought, only 1 pound of water is available,

$$\frac{1}{18} = \frac{z}{98}$$
$$z = 5.4 \text{ pounds } H_2SO_4)$$

All chemical reactions use and produce exact amounts of materials in this way. If the equation is accurate, these amounts can be calculated from the equation alone.

A beginner in chemistry might understand the preceding equation well enough to see that $SO_3$ and water produce sulfuric acid but not understand that the amounts involved are determined by the molecular weights. He might then arbitrarily combine equal amounts of $SO_3$ and water. If so, only ¼ of the $SO_3$ would have enough water to react with; the remaining ¾ would go up the chimney.

Here are five common types of chemical reaction:

(1) *Synthesis*. A direct combining of two or more reactants. The formation of $H_2SO_4$ is an example. So is the rusting of iron:

$$4Fe + 3O_2 \rightarrow 2Fe_2O_3$$

(2) *Decomposition*. A breaking down of a compound into two or more products. For example, what happens when a bottle of beer is opened:

$$H_2CO_3 \rightarrow H_2O + CO_2\uparrow$$

(3) *Displacement*. What occurs when a more active element pushes a less active element out of a molecule:

$$Zn + H_2SO_4 \rightarrow H_2\uparrow + ZnSO_4$$
(zinc sulfate)

(4) *Double Displacement.* What occurs when two compounds form two new compounds by exchanging portions of their molecules. One pair is usually more strongly drawn together than the other. In the following examples these are represented by AgCl and HOH:

$$AgNO_3 + NaCl \rightarrow AgCl\downarrow + NaNO_3$$
(silver nitrate) (sodium chloride) (silver chloride) (sodium nitrate)

$$KOH + HCl \rightarrow KCl + HOH$$

(5) *Reduction-Oxidation* (*Redox*). Originally oxidation meant "combining with oxygen" and reduction meant "combining with hydrogen." Then it was found that the significant part of the reactions involved the loss and gain of electrons, and that this could happen without the direct involvement of oxygen or hydrogen. The terms were redefined accordingly.

There can be no oxidation without a simultaneous reduction, and no reduction without a simultaneous oxidation. Electrons can only be "gained" after they have been "lost" from something.

Electrons, as is explained in chapter 16, are mobile particles on the outermost portions of atoms. Since electrons bear a negative (−) charge, their departure leaves a residual positive (+) charge.

When a piece of iron is placed in a solution of copper sulfate

$$CuSO_4 \rightarrow Cu^{++} + SO_4^{=}$$

the iron dissolves, while the copper plates out:

$$Fe \rightarrow Fe^{++} + 2 \text{ electrons} \quad \text{(oxidation)}$$
$$2 \text{ electrons} + Cu^{++} \rightarrow Cu \quad \text{(reduction)}$$

$$Fe + Cu^{++} \rightarrow Fe^{++} + Cu \quad \text{(the redox reaction)}$$

Some chemical reactions give off energy (explosions, burning of fuels), while others require energy (heat, sunlight, electricity) in order to take place. Some reactions are reversible, others are not. For most reactions, the energy needed or released, the rates, and the final concentrations of all reactants and products can be calculated.

The easiest way to make a reaction "go" is to remove one of the products. If beer is boiled, driving off the $CO_2$, it would be very hard to recapture that $CO_2$ and put it back in the beer. In a neutralization reaction, the formation of water effectively removes the $H^+$ and $OH^-$ ions from the solution. Formation of an insoluble compound has a similar effect; formation of AgCl (see reaction type 4) removes both $Ag^+$ and $Cl^-$ ions from the reaction.

All these conditions make a reaction "go to completion," with little tendency to reverse itself.

The next step is to learn why certain compounds form readily, while others do not. For this more information about the inner structure of atoms is required.

# 16. The Periodic Table

*Ah, yes. Arranging and collecting—that's what you're so good at.*
——Henrik Ibsen, *Hedda Gabler*

During the age of exploration, European ships returned from all over the world with quantities of plant and animal specimens, most of them new and unfamiliar. The staffs of the museums which ultimately received this material had to find some way of organizing it, and as a result, singularly efficient classification systems were developed. Something similar happened during the nineteenth and twentieth centuries, when the 33 elements in Lavoisier's *Traité* grew to over 100.

A beginning botany student does not learn the names and characteristics of all 200,000 species of flowering plants. He separates them into a few broad divisions: those with flowers above the seeds (apple and dogwood), those with flowers below the seeds (lotus and magnolia), and so forth. Then he goes into as much additional detail as his needs and interests require.

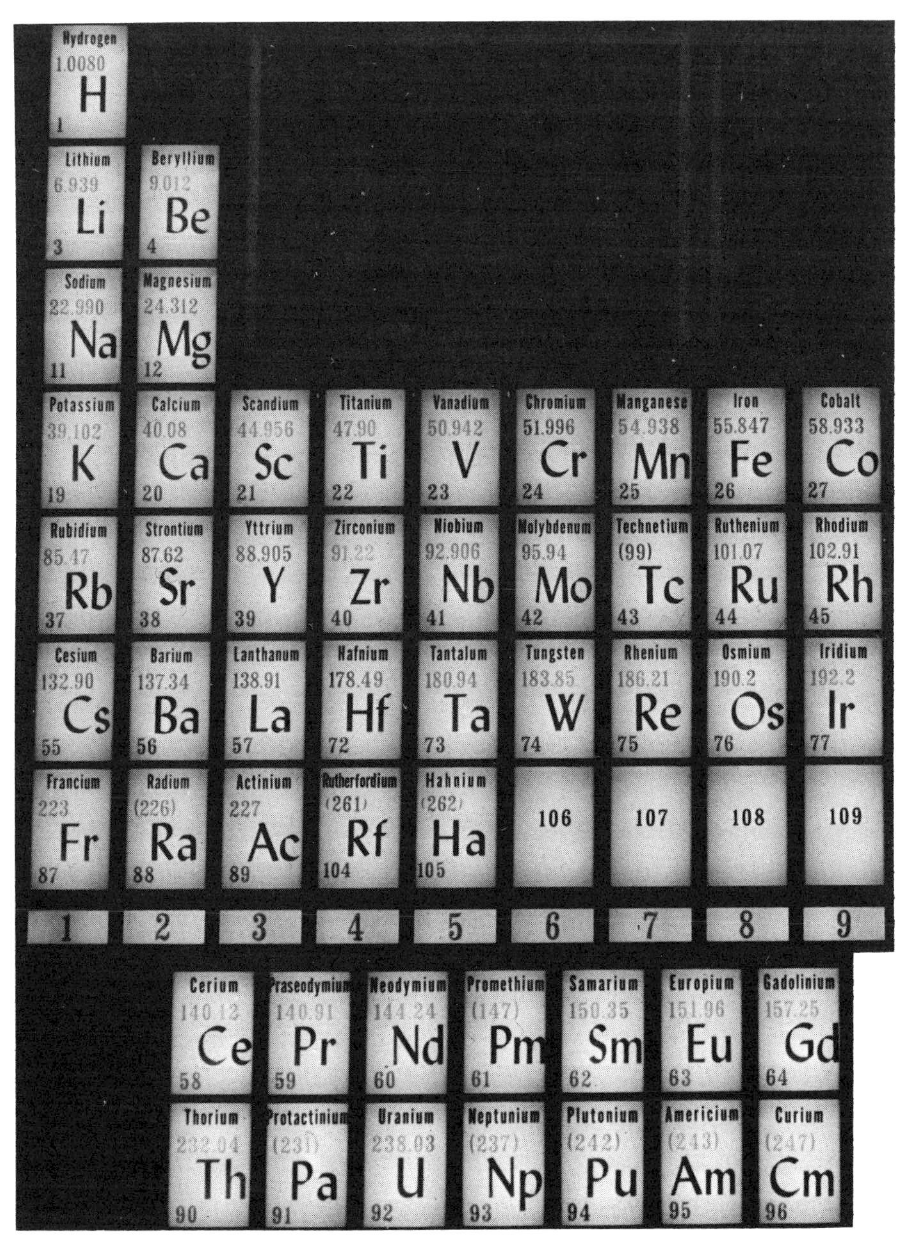


FAMILIES OF ELEMENTS REPRESENTED BY NUMBERED COLUMNS

| | | | |
|---|---|---|---|
| I A | 1 | V A | 15 |
| II A | 2 | VI A | 16 |
| I B – VIII B | 3–12 | VII A | 17 |
| III A | 13 | VIII A | 18 |
| IV A | 14 | | |

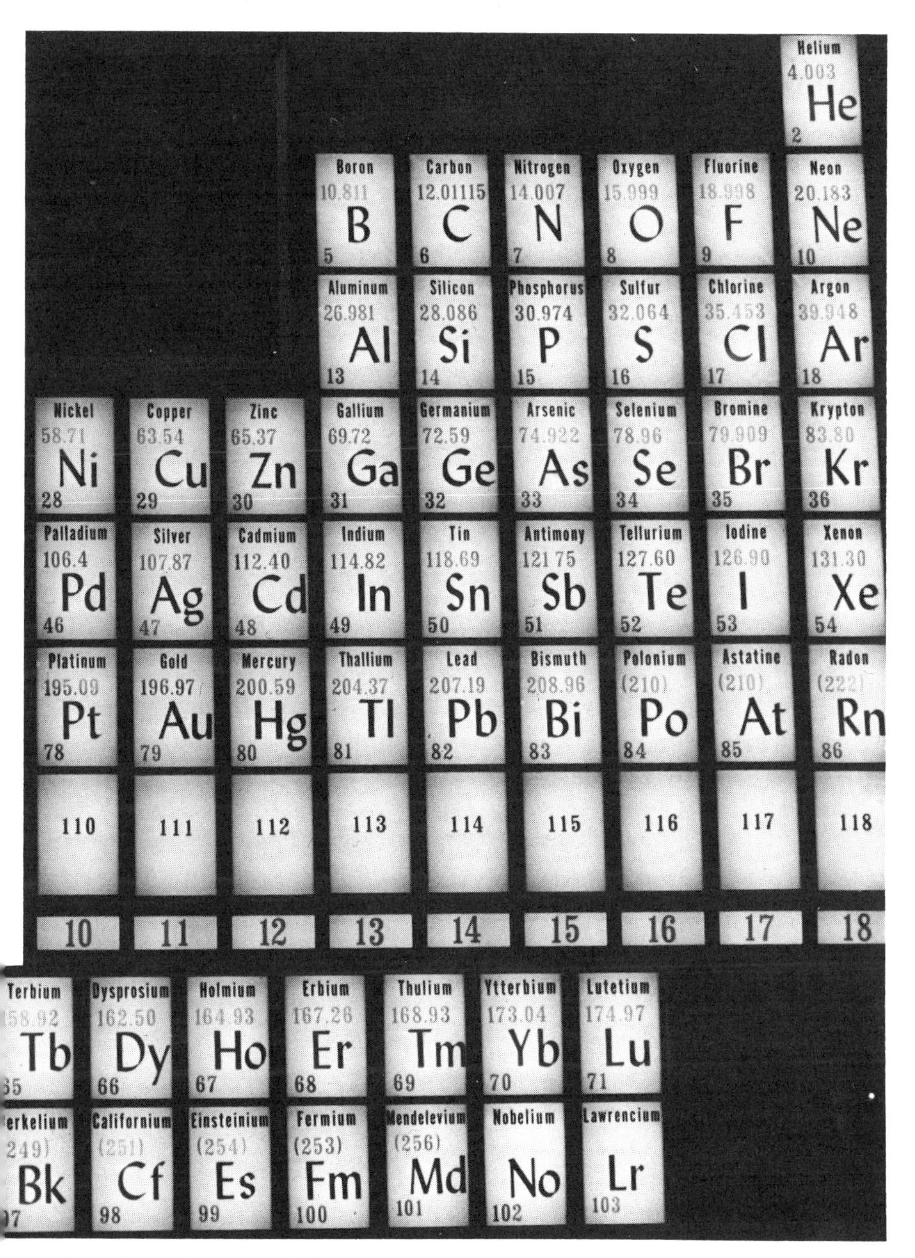


Periodic table designed by Harvey E. White and the staff of the Lawrence Hall of Science, Berkeley, California

This is taxonomy, the science of classification. Chemical taxonomy involves the Periodic Table of the Elements (see preceding pages). Besides its organizational value in allowing chemists to deal with groups of related elements, this table has stimulated research in atomic structure and helped explain the results.

Some elements obviously resemble one another. Metals are similar to one another and different from nonmetals. As early as 1829, Goethe's chemistry teacher at Jena, Johann Döbereiner (1780–1849), was arranging similar elements into "triads." The alkali metals, lithium (Li), sodium (Na), and potassium (K) are all soft, light-weight, and violently reactive. Chlorine (Cl), bromine (Br), and iodine (I) all have sharp odors, form salts easily, and produce similar acids with hydrogen (HCl, HBr, HI). The alkaline earth metals, calcium (Ca), strontium (Sr), and barium (Ba), look and react alike in many ways.

An English chemist, John A. R. Newlands (1837–1898), listed the elements in order of increasing atomic weight in eight vertical rows and noticed that the properties of every eighth element seemed to be similar, like every eighth note in a musical scale. Newlands's "law of octaves" sent the London Chemical Society into gales of laughter in 1865, but in 1887 he was awarded the Davy Medal by the Royal Society.

There was no place for additional elements in Newlands's system; when a new element was discovered, the system had to be revised. A more flexible and workable scheme was developed by the Russian chemist Dmitri Mendeleev (1834–1907) in 1869. In Germany Lothar Meyer developed a similar but less complete system independently.

The youngest of seventeen children, Mendeleev was born in Tobolsk, Siberia. His remarkable mother supported the family by managing a glass factory. But when her blinded husband died and the factory burned down, she took Dmitri (the most talented of her brood) several thousand miles to Saint Petersburg and got him admitted to the University.

Mendeleev received his doctorate and won a fellowship at Heidelberg, then became a professor at the University of St. Petersburg. His liberal views kept him at odds with the government (in czarist Russia at that time one could get a reputation for being "liberal" simply by playing cards on Sunday). Once a year, in spring, he had his luxuriant hair and beard trimmed; he saw no reason to change this schedule, even when he was presented to the Czar. He received many foreign honors, but he was never elected to the Imperial Academy of Sciences of St. Petersburg.

His filing system helped him develop the Periodic Table. He put the name of each known element on a card, along with all of its properties. When he arranged the cards in order of increasing atomic weight, similar properties did indeed recur at regular intervals.

He put the cards in horizontal rows in such a way that each vertical row contained the related elements. If an element did not seem to fit in a given space, he left the space blank, correctly assuming that an element not yet discovered belonged there. His success at predicting the existence of unknown elements was a major factor in winning acceptance for his system.

Many of today's Periodic Charts are overwhelmingly detailed, so crammed with information that a novice can stare at one fifteen minutes without even finding the element he wants. But most of the basic relationships shown by Mendeleev are still present. To bring this out, I have recast part of Mendeleev's original chart in the modern format. All these elements and their positions are exactly as proposed by Mendeleev in 1869:

| *I A* | *II A* | *III A* | *III B* | *IV A* | *V A* | *VI A* | *VII A* |
|---|---|---|---|---|---|---|---|
| Li | – | B | | C | N | O | F |
| Na | – | Al | | Si | P | S | Cl |
| K | Ca | ?–1 | ?–2 | ?–3 | As | Se | Br |
| Rb | Sr | – | – | Sn | Sb | Te | I |
| Cs | Ba | – | – | – | Bi | – | – |

The Roman numerals are post-Mendeleev and indicate "families" of elements:

| | |
|---|---|
| I A | alkali metals |
| II A | alkaline earth metals |
| III A | boron family |
| IV A | carbon family |
| V A | nitrogen family |
| VI A | oxygen family |
| VII A | halogens |
| VIII A | inert gases (not discovered until the 1890s) |
| I B–VIII B | transition metals |

Mendeleev predicted that one of the blank spaces on the original chart, "?–1," would be filled by a metallic element with a low melting point, an atomic weight near 68, and an oxide with the formula $X_2O_3$ (X meaning the unknown element). Six years later (1875) a metal was discovered that melted at 30° C., had an atomic weight of 69.7, and formed the predicted oxide ($Ga_2O_3$). Its discoverer was a patriotic Frenchman, Paul de Boisbaudran, who named the element gallium (Ga) after the Latin name for France, Gallia.

In 1879 a patriotic Scandinavian, the Swedish chemist Lars Fredrik Nilson, discovered an element that fitted blank space "?–2" and named it scandium (Sc). And in 1886 a patriotic German, Clemens Winkler, discovered "?–3," which he called germanium (Ge).

Local pride has continued to influence the naming of elements: polonium (1898), francium (1939), americium (1944), berkelium (1949), and californium (1950).

Not all of Mendeleev's predictions worked out so well, but the three successes within twenty years were enough to convince most scientists of the validity of his system.

The periodic law was accepted and used, even though there was no theoretical explanation as to why it worked. That came later.

# 17. The Anatomy of Atoms

THE DEVIL: *Don Juan, shall I be frank with you?*
DON JUAN: *Were you not so before?*
THE DEVIL: *As far as I went, yes. But I will now go further. . . .*

——George Bernard Shaw, *Man and Superman*

Often scientists find themselves forced to abandon a position they have just proclaimed to be irrevocably true. Like the wise man Kamadamana in Thomas Mann's *The Transposed Heads,* they are reduced to insisting, "What I said last—that goes."

After proving that life cannot arise spontaneously, biologists now collect evidence to show that it can. After convincing the world that changing one element into another was a heretical alchemical idea and that it could not possibly be done, chemists are now changing one element into another.

Small wonder that so many older scientists hate to make unqualified statements on any subject. But if the aim of science is truth, the craving for infallibility must always be sacrificed to new information.

Nineteenth-century chemistry was based firmly on Daltonian atoms—those tiny indestructible billiard balls. After the 1860 Karlsruhe congress, the pace of achievement quickened: pioneering discoveries were made in biochemistry, the syntheses of organic compounds multiplied incredibly, the great chemical industries burgeoned. Many chemists now enjoyed new levels of prestige, wealth, and power.

But by the end of the century it was becoming clear that the Daltonian atom was finished. The concept failed to explain the newer experimental evidence about atomic structure. Far from being indestructible, atoms now appeared to be full of holes and might fall to pieces spontaneously. Moreover, some of the pieces had surprising, even alarming properties.

These discoveries were not made by chemists, but by physicists, most of them English. Less worldly as a rule than the chemists, they carried on a quiet, relentless investigation into the fundamental nature of matter that has succeeded beyond all expectation.

Chemists have taken varying attitudes toward the contributions of physicists. Sometimes they blur the distinctions and mentally convert the physicists into chemists, or convert themselves into physicists, or pretend that the remarkable discoveries came from chemists and physicists "working hand-in-hand." Nearly always they praise the physicists. But no matter how warm the praise, one senses an infinite regret that at such a critical moment in the race the baton was snatched from their hands.

## The Cathode Rays

From 1785 on, physicists studied the passage of electric currents through gases at low pressures. The terminals were usually sealed at opposite ends of a glass tube:

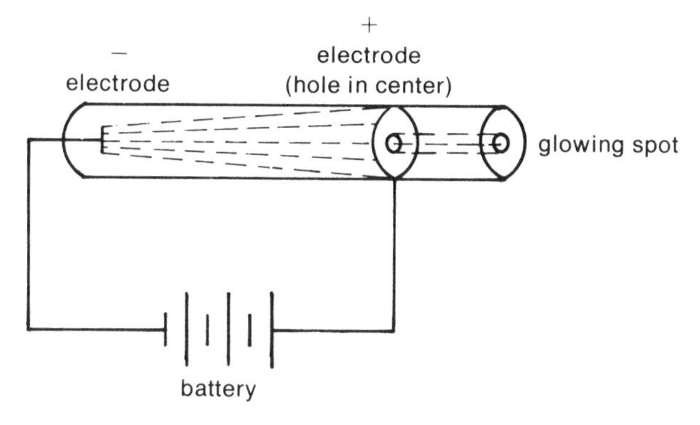


The glass near the cathode (the negative terminal or electrode) was often observed to glow. Magnets caused the glowing area to shift position. It was concluded that the cathode was emitting some kind of "rays."

The English chemist and physicist Sir William Crookes (1832–1919) removed air from his tubes, down to pressures of 14/1,000,000 mm. Hg, thus inventing the Crookes tube. After several years of study, he suggested that cathode rays were electrified particles. In Germany, Heinrich Hertz (1857–1894), who had discovered radio waves, preferred to think of cathode rays as "waves." (It has been suggested that love of ball games such as cricket makes the English favor "particle" theories, while love of music makes the Germans favor "wave" theories.)

Cathode rays could be deflected by the negative pole of a

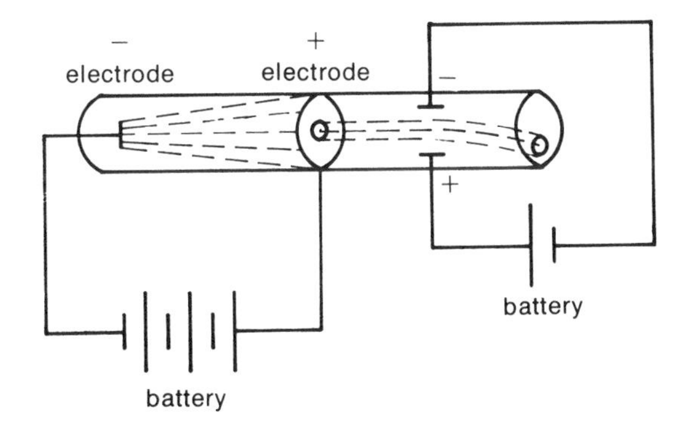

magnet, showing that they were negatively charged. The great English physicist Sir Joseph John Thomson (1856–1940) also deflected them with electric fields and showed unequivocally that cathode rays were streams of negatively charged particles, each having $^{1}/_{1836}$ the weight of a hydrogen atom.

The particles were called "electrons," from the Greek word for amber, a source of negative electricity.

## X Rays and Radioactivity

In 1895, the German physics professor Wilhelm Röntgen (1845–1923) accidentally found that cathode rays could give rise to a far more powerful component. When cathode rays struck glass (later, metals were used), invisible rays were generated which passed through wood, aluminum sheets, flesh, affected photographic film, were partially stopped by bone, and entirely stopped by lead.

He called them "X rays," and his discovery received wide publicity, much of it unbelievably silly. Apparently in 1895 (the year that Sigmund Freud and Josef Breuer published their *Studies in Hysteria*) the idea of seeing *anything* below the surface was highly upsetting. Nothing could convince the press that the sole purpose of X rays was not to peek at the rosy flesh of unsuspecting citizens. In New Jersey a law was introduced forbidding the use of "X-ray opera glasses." In London X-ray-proof undergarments were offered for sale.

One good effect of the publicity was that it attracted the attention of a French physicist, Antoine Becquerel (1852–1908). He had been working with a collection of uranium salts left him by his father. If exposed to sunlight and then taken into a dark room, the salts would glow. After reading about Röntgen, Becquerel wondered if his uranium salts might not be emitting X rays.

The uranium salts fogged up photographic film whether

they were exposed to sunlight first or not. He decided that the uranium was emitting something even better than X-rays, something which ought to be called "Becquerel rays," in fact. He then turned the problem over to one of his graduate students.

The student, Marya Sklodovska (1867–1934), better known as Marie Curie, had fled to Paris in 1891 from a Poland controlled by czarist Russia. Strong-willed and industrious, she had worked as a bottle washer while taking courses in physics and mathematics at the Sorbonne. In 1894 she met a professor Pierre Curie, a confirmed bachelor in his late thirties. Marie liked him. Within a year they were married. Since neither was especially religious, the wedding was a simple civil ceremony. Marie did not even bother to change her dress. They lived for a while in a three-room flat where, for lack of funds, the only furniture was a bed. Neither complained.

Working for her doctorate now, Marie studied a number of radioactive materials. She found that one of Becquerel's specimens of uranium ore was more active than pure uranium compounds; this strongly suggested that the ore contained something more radioactive than uranium itself. At her request, the Austrian government sent her a ton of pitchblende from which the uranium had been extracted. For the next two years she and Pierre worked at concentrating this great mass of material.

Fumes from the boiling vats in the laboratory nearly killed them, so Marie carried the vats out to the yard and continued there. During the freezing winter, she caught pneumonia, then she had a baby, but nothing slowed her down for long. By the time she had concentrated the ore down to 100 pounds, both she and her husband were sick much of the time. However, she later looked back on this as the happiest period of her life.

When she finally crystallized the salt of the new elment, it proved to be 300 times more radioactive than uranium. She named it polonium after her homeland. Visitors at this time said she would only talk on three subjects: science, Poland, and the baby.

She detected a third substance in the ore concentrates and worked at isolating this. About this time the Curies noticed that the tubes and bottles in the laboratory at night glowed in the dark and gave off heat. At last she obtained the salt of another new element, millions of times stronger than uranium: radium.

No original methods, no ingenious new chemical techniques, were used in any of this work. What makes the achievement unique is the sheer physical energy, the will power, and the "terrible patience." Marie was now cooking meals, making dresses for the baby, and studying for her oral examinations at the Sorbonne. Her doctoral thesis, *Recherches sur les Substances Radioactives,* was published in 1903. That same year she shared the Nobel Prize for Physics with her husband and Becquerel for their work in radioactivity.

A few years later, Pierre Curie was killed by a truck. Marie took over his teaching position. In 1911 she received a second Nobel Prize, this time in chemistry, for the isolation of metallic radium.

Radium is poisonous. A tiny grain of it kills a mouse in half a day. Pierre's hands became half-paralyzed from handling it. Becquerel burned his abdomen by carrying a vial of radium to his vest pocket for a lecture in London. In spite of the danger, the element was intensively studied, especially its radiation.

It was quickly established that part of the radiation consisted of particles just like the cathode-ray particles. In addition, there was a highly penetrating radiation, very much like X rays, and this was the component that did most of the damage.

One of J. J. Thomson's students, Ernest Rutherford (1871–1937), gave the names alpha ($\alpha$), beta ($\beta$), and gamma ($\gamma$) to the three kinds of radiation that eventually were ascribed to radium (see illustration on following page).

Rutherford also was quick to appreciate that the particles being emitted were *parts of the atoms* of radium. The expulsion of these particles might leave behind atoms that were "lighter than before and possessing physical and chemical properties

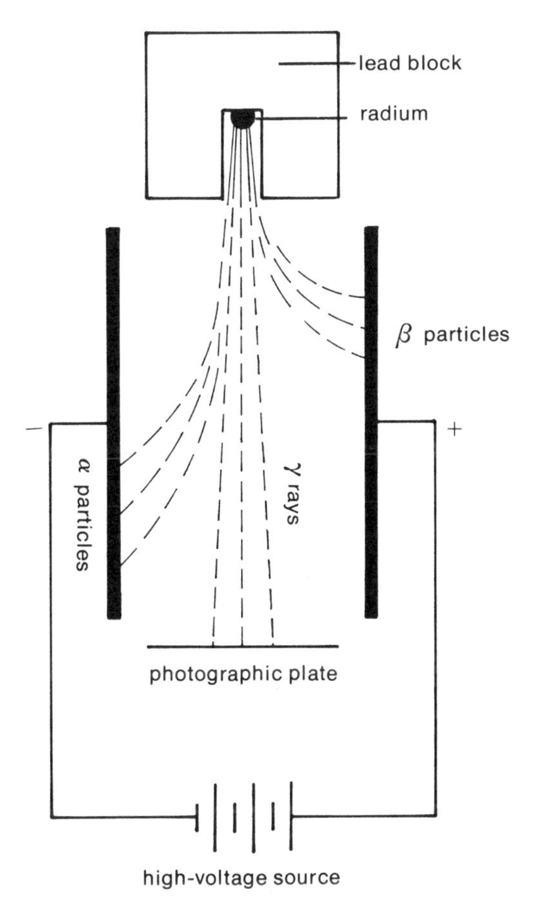


quite different from those of the original parent element," he and his colleague Frederick Soddy (1877–1956) reported in *Philosophical Magazine* in 1902. In other words, the radium atoms might be turning into the atoms of some other element.

The various stages of disintegration were worked out by which uranium turned successively into thorium, radium, radon, polonium, lead, and bismuth, each with a different half-life. Rutherford identified the alpha particle as a positively charged helium atom, and his experiments with it led him to give the first modern description of atoms in general.

## The New Atom

According to Rutherford (1911), an atom consisted of a hard positively charged core around which negatively charged electrons revolved, like planets around the sun. In 1912, he named this central core of the atom the "nucleus."

The simplest element, hydrogen, had a single positively charged particle as its nucleus, with a single electron revolving around it. In 1914 Rutherford named this positively charged hydrogen nucleus a "proton." It corresponds to the hydrogen ions ($H^+$) released by acids in water solution:

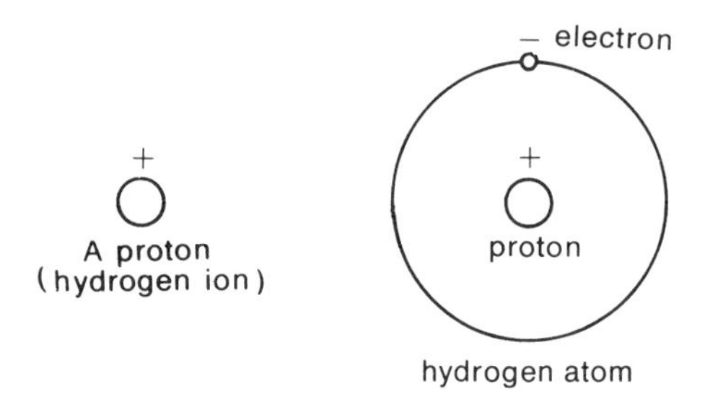


One of Rutherford's students, Henry Moseley (1887–1915), studied the nuclear charge of various elements. He found that each successive element in the periodic table gained one more positive charge (proton). The elements in the periodic system had been numbered in order of increasing atomic weight; these "atomic numbers," Moseley discovered, correspond exactly to the number of positive charges (number of protons) in the atomic nucleus. And since the atom is electrically neutral, the atomic number also corresponds to the number of electrons revolving around it. Thus H=1, He=2, Li=3, and so on.

Moseley showed that the chemical properties of an element depend on its number of electrons, not on its atomic weight. An atom with 7 electrons (and of course 7 protons) is nitrogen; one with 8 electrons, oxygen; 9 electrons, fluorine; and 10 electrons, neon.

Yet atomic weight continued to present problems. Hydrogen, with one proton and one electron, had an atomic weight of 1. Helium, with two protons and two electrons, had an atomic weight of 4. The explanation was that there was a third kind of particle in the nucleus: the neutron. It had about the same weight as a proton, but no electrical charge. It affected mass only, not chemical properties. This explained the occurrence of isotopes (from Greek *isos,* same; *topos,* place). Isotopes are atoms of the same element having different numbers of neutrons. The hydrogen isotopes, deuterium and tritium, behave like hydrogen, but weigh more:

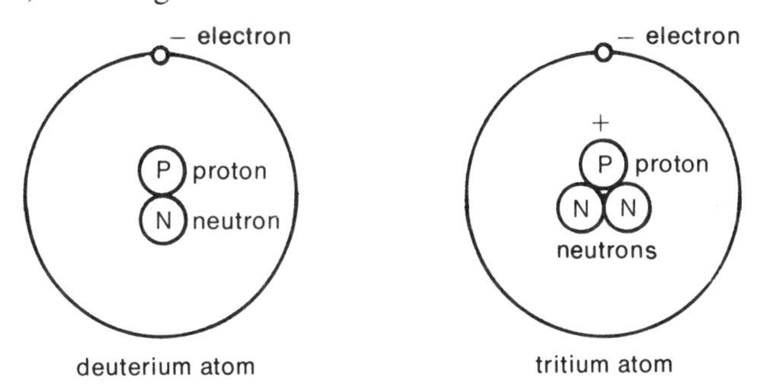


This is one way of "labeling" a compound. The fate of the "heavy" molecules can be followed in living organisms, or leaky pipes, and they can always be detected because of their abnormally high weights.

Another Rutherford protégé, the Danish physicist Niels Bohr (1885–1962), modified the new conception of the atom so as to agree with the quantum theory of the German physicist Max Planck (1858–1947). This theory forms the basis for the atomic physicist's view of atoms. Chemists, however, still use the more naive atom picture developed by Gilbert Lewis (1875–1946), an American professor of physical chemistry, and extended by Irving Langmuir (1881–1957), a physicist for General Electric Company. It may make physicists wince, but it is supremely useful in ordinary chemical practice. It is described in the following section.

## A Second Look at the Periodic Table

As the number of electrons in the elements increases, they do not all move around in the same orbit but are grouped in a series of concentric orbits. Only the outer ring is involved in ordinary chemical reactions. As might be expected from the eight groups in the Periodic Table and from Newlands's "octaves," the outer ring of most elements can hold a maximum of 8 electrons, and this is its condition of maximum stability.

In the Lewis-Langmuir system, the 8 potential electron spaces are at the corners of an imaginary cube around the nucleus. The Roman numerals for the groups in the Periodic Table correspond to the numbers of electrons actually occupying the potential spaces. Group I elements have just 1 electron in the outer shell, group IV elements have 4 electrons, group VII elements have 7 electrons, and so on.

Since the group VIII A elements have all 8 spaces filled with electrons, they have already achieved maximum stability. They are, in fact, the inert gases. For the other groups, stability can be achieved by one of three strategies:

(1) giving up all the electrons of the outer shell, so that the (completed) shell underneath now becomes the outside shell
(2) obtaining electrons from another atom to complete the outer shell
(3) sharing electrons with another atom.

Atoms having 1 or 2 electrons can give these up more easily than they can acquire 6 or 7 electrons from other atoms. Atoms with 6 or 7 electrons can pick up 1 or 2 more electrons more easily than they can unload the entire outer shell. Atoms with 4 electrons in the outer shell are evenly balanced between losing and gaining electrons.

Electron losers are located on the left-hand side and in the middle of the Periodic Table. They are metals. The more active metals are toward the far left (fewer electrons to lose) and toward the bottom (the higher the atomic weight, the farther the electrons are from the nucleus and hence the more easily electrons are lost).

Electron gainers are located on the right-hand side of the Periodic Chart. They are nonmetals. The more active nonmetals are at the far right (fewer electron vacancies to fill) and at the top (the lower the atomic weight, the closer the electron vacancy is to the nucleus and the stronger is the force of attraction, as with gravity).

By now it should be clear that loss or gain of electrons is the basis of all chemical reactions.

Chemical change can be seen as an attempt of the atoms involved to achieve maximum stability. NaCl is more stable than either Na or Cl:

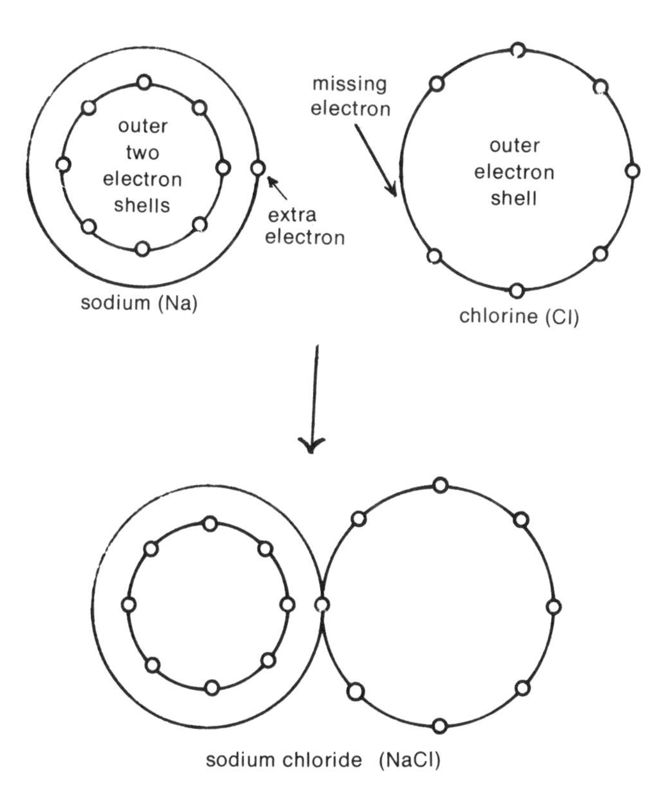

Similarly, $Na^+$ ions and $Cl^-$ ions are more stable than Na or Cl:

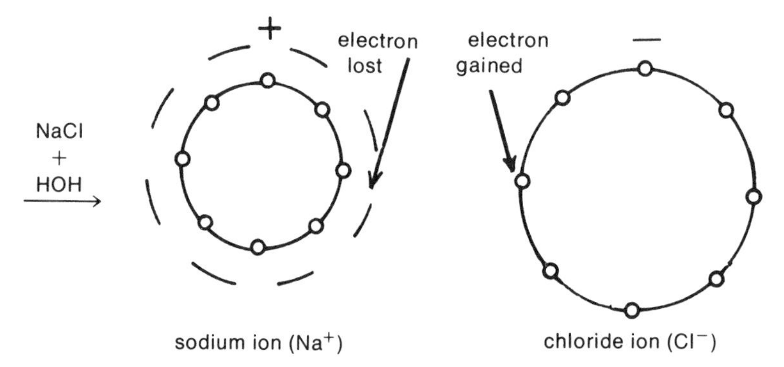


At the higher atomic weights the electron shells build up in a more complicated fashion, and the Lewis-Langmuir system is less successful. Also, the very first electron shell can hold only 2 electrons, which means helium is stable, but hydrogen is not. (Stability is a full electron shell. If the electron shell can hold only 2 electrons, rather than 8, then helium with 2 electrons has maximum stability.)

## Sunya

A tradition in Mahayana Buddhism holds that the physical world (*rupa*) is actually the void (*sunya*). The Indian sage Nagarjuna advanced this "doctrine of the void" in the 2nd or 3rd century A.D.

In one of Rutherford's experiments, a thin stream of alpha particles from radium in a lead container was directed toward a piece of gold foil 2,000 atoms thick. A spherical fluorescent screen enclosed the foil entirely except for the hole admitting the stream of alpha particles; when a particle struck the screen, a tiny flash of light was produced. Out of 100,000 heavy alpha particles, 99,999 went right through the gold foil; only one particle struck something in the foil that deflected it backward.

In other words, the gold foil may appear solid, but there is much more "sunya" in it than solid material. A rough idea of how much of an atom is empty space can be gained by imagining a single BB-gun pellet in the middle of the Houston Astrodome. The "BB" represents the nucleus of a hydrogen atom. The outer limit of the Astrodome represents the orbit of the revolving electron.

So an atom is mostly empty space, like the solar system. If someone fired a succession of missiles—at random—at the solar system, the chances of their hitting anything would be very slight.

# 18. The Halogens

The elements of the halogen family are usually the first to be studied after the Periodic Table has been introduced. Since the halogens are clearly related to one another, they seem to offer overwhelming proof of the table's logic and correctness. Other families, in which the relationships are less apparent, are presented later, less obtrusively.

Halogen means "salt former." Each halogen atom has 7 electrons in its outer shell, with room for one more. Since metals lose electrons easily, they readily form salts with the halogens. The best-known salt of all, ordinary table salt ($NaCl$) itself, is one of these. (See table on following page.)

The higher the atomic weight, the more electron shells, and the farther the outer shell is from the nucleus. In fluorine and chlorine, the electron vacancy of the outer shell is close to the positively charged nucleus and able to exercise great attractive force. Chemical reactivity of the halogens (and of nonmetals generally) goes down as the atomic weight increases.

| *Element* | *Atomic weight* | *Physical state* | *Number of electron shells* |
|---|---|---|---|
| fluorine (F) | 19 | gas | 2 |
| chlorine (Cl) | 35 | gas | 3 |
| bromine (Br) | 80 | liquid | 4 |
| iodine (I) | 127 | solid | 5 |
| astatine (At) | 210 | solid | 6 |

*Fluorine* (*F*). A pale yellow gas, fluorine is never found free under natural conditions. It forms fluorides with metals, such as sodium fluoride (NaF). Many of these occur in common rocks and minerals (fluorspar, $CaF_2$).

Fluorine reacts with nearly all of the elements, often violently and with spontaneous ignition. This characteristic made it very hard to isolate. Whenever free fluorine was liberated in the laboratory, it immediately combined with whatever was around it: the reaction vessel, a chemist's flesh. Reaction vessels of fluorspar or copper are now used. With copper, the copper fluoride that is instantly formed acts as a protective layer.

Fluorides from dissolved minerals are found in small amounts in most natural waters. They are toxic in excess, but traces of fluoride are necessary to life and for the formation of durable tooth enamel. Sometimes it is added to water supplies that lack it at a rate of 1 part per million (ppm). Natural waters containing 6 ppm or more may cause a mottling of the teeth.

Hydrogen reacts with especial violence with fluorine to form hydrofluoric acid:

$$H_2 + F_2 \rightarrow 2HF$$

Hydrogen also reacts, although with increasing sluggishness, with the other halogens:

$$H_2 + Cl_2 \rightarrow 2HCl \quad \text{(hydrochloric acid)}$$
$$H_2 + Br_2 \rightarrow 2HBr \quad \text{(hydrobromic acid)}$$
$$H_2 + I_2 \rightarrow 2HI \quad \text{(hydroiodic acid)}$$

Hydrofluoric acid is poisonous and dissolves glass. It is used to "frost" electric light bulbs and to etch designs or letters on glass. Needless to say, it is not stored in glass bottles; polyethylene bottles or bottles lined with wax are used.

Fluorocarbon plastics are highly inert. "Teflon" is used to make laboratory ware and nonsticking cooking utensils.

"Freon" ($CCl_2F_2$) is the refrigerant in most household refrigerators and the propellant in many aerosol sprays.

*Chlorine* (*Cl*). This irritating, greenish-yellow gas was the first poison gas to be used in modern warfare (1915). Aroused world opinion stopped the use of this weapon after World War I.

Chlorine is the most abundant of the halogens. Over 6 million tons of it are produced yearly in the United States. It is used to bleach "newsprint," to purify drinking water, and to synthesize a number of dry-cleaning solvents.

"Salt," sodium chloride (NaCl), is the most abundant and widely used halogen compound. Underground deposits of it in some areas can be mined like coal. Before modern refrigeration, it was widely used to preserve food. Primitive people valued it and used it as money. Wars were frequently fought over the control of salt deposits.

Sodium hypochlorite (NaClO) is a strong oxidizing agent and a commonly used laundry bleach (5-percent solution). A comparison of the cost of a nationally advertised brand with that of a "Co-op" or chain-store brand provides a rough idea of the cost of advertising.

The bottles of "concentrated HCl" seen in laboratories usually contain 40 percent HCl in water. Pure HCl is a colorless gas.

*Bromine* (*Br*). A reddish-brown fuming corrosive liquid with a strong smell (*bromos,* stench), bromine is not an abundant element. It is usually extracted from sea water.

The tetraethyl lead added to gasoline since 1924 will foul

ignition points unless ethylene dibromide ($C_2H_4Br_2$) is also added. This permits the lead to escape through the exhaust as the toxic and irritating gas $PbBr_2$.

Potassium bromide (KBr) and other bromides were used as tranquilizers at the end of the nineteenth century. Anton Chekhov mentions potassium bromide in *The Black Monk* (1894). It is still an ingredient in Bromo-Seltzer and various "nerve tonics," despite its potentially toxic side effects.

Silver bromide (AgBr) is one of the light-sensitive materials used in photographic film and enlargement paper.

*Iodine* (*I*). When heated, the lustrous, gray-black crystals of iodine form a violet vapor, instead of melting; the vapor recrystallizes upon cooling. Like carbon dioxide, iodine normally skips the liquid phase.

Iodine occurs in very small amounts in the ocean, but seaweeds concentrate it. It is necessary for animal life, especially for the normal functioning of the thyroid gland. A human requires 0.0001 gram of iodine each day. One gram (0.035 ounce) would last him twenty-four years.

Natives of areas of the midwestern United States where iodine is deficient used to develop goiters (enlargement of the thyroid gland). "Iodized" salt (NaCl plus a small amount of NaI) prevents this condition.

*Astatine* (*At*). Traces of astatine (*astatos,* unstable) may be found in uranium ores. It is created artificially by bombarding bismuth (Bi) with alpha particles.

This element is radioactive and short-lived, lasting only about 8 hours. Fast-working chemists have established that it generally resembles iodine and forms salts (and therefore can be kept in the halogen group) but also has some unique reactions.

# 19. The Nonmetals

The upper far-right portion of the Periodic Table is occupied by the nonmetallic elements. A line running diagonally in steps from just below boron to astatine separates the official metals from the official nonmetals.

Once a line is drawn or a rule is made, there is an immediate rush to change it. Elements near the dividing line have both metallic and nonmetallic properties. By calling these elements "metalloids," one is spared the embarrassment of describing a silvery element which conducts electricity (tellurium) as a nonmetal. Unfortunately, different countries and different professions have their own ideas as to which elements are metalloids, and this naturally affects which elements they call "true" metals and "true" nonmetals.

An objective way of resolving the problem is to use the electron-attracting power of an element as the criterion. Metals are good conductors, nonmetals are good insulators, elements halfway between are metalloids. This criterion gives five metalloids: boron, silicon, arsenic, selenium, and tellurium.

Carbon is equivocal, since in one form (diamonds) it is an insulator, while in another (graphite) it is a conductor. Since carbon is unique and extremely important, it is given special consideration in chapter 21.

The following nonmetals are left: the inert gases, the halogens, oxygen, nitrogen, sulfur, and phosphorus.

Hydrogen is another special case: the source of the other elements. It is obviously nonmetallic, however, and sometimes is placed with the halogens (or, along with the alkali metals, above the dividing line).

The only nonmetals not already discussed are sulfur and phosphorus.

*Sulfur (S).* This is the second element in group VI A of the Periodic Table. As a bright yellow powdery solid, sulfur does not resemble the element above it (the colorless gas oxygen) or the elements below it (the metalloids selenium and tellurium). Chemically, it bears some resemblance to oxygen.

When the Bible mentions brimstone (a stone that burns), sulfur is meant. It is found around volcanoes in Japan and Mexico and is especially common in the Mediterranean (the islands of Sicily, Stromboli, and Vulcano). Sulfur burns with an eerie blue flame and a bad smell, very suitable for hell.

Until 1903, 95 percent of the world's sulfur came from Sicily. The monopoly was broken by the German-born American chemist Herman Frasch (1851–1914).

Huge deposits of sulfur had been found in eastern Texas and western Louisiana, but they were buried under several thousand feet of quicksand. Conventional shaft mining was out of the question. Frasch developed an ingenious process for getting the sulfur to the surface; engineers ridiculed the process for eight years before he was given the chance to demonstrate it in 1894.

He drove three concentric pipes down into the sulfur deposit. Superheated water (180° C.) was forced down through the outer ring; compressed air was forced down through the center pipe. Molten sulfur and hot water were thus forced up the pipe through the middle ring and dumped into storage tanks. The water drained away, leaving behind 99.9 percent pure sulfur.

The Frasch process is still widely used. Some wells produce 500 tons of sulfur per day. Since it is not very reactive at ordinary temperatures, sulfur can be left in outdoor storage bins.

At higher temperatures, however, sulfur burns, producing the irritating and strong-smelling gas sulfur dioxide:

$$S + O_2 \rightarrow SO_2\uparrow$$

The Mauna Loa volcano on the island of Hawaii emits large amounts of sulfur dioxide. With water, $SO_2$ forms a weak, unstable acid:

$$H_2O + SO_2 \rightleftharpoons H_2SO_3 \text{ (sulfurous acid)}$$

Sulfur dioxide is a good bleach for silk, wool, and cellulose. Its effects are not always permanent, however. Chips of wood cooked with $SO_2$ and $Ca(OH)_2$ turn into a "sulfite pulp" of cellulose, which is used for books and newspapers.

With more oxygen and a catalyst (such as vanadium pentoxide, $V_2O_5$), sulfur dioxide becomes sulfur trioxide:

$$2SO_2 + O_2 \rightarrow 2SO_3$$

Sulfur trioxide is an extremely irritating, choking gas, a key ingredient in the "killer smogs" of London and Donora, Pennsylvania. With water, $SO_3$ forms one of the strongest and most corrosive acids known:

$$H_2O + SO_3 \rightarrow H_2SO_4 \text{ (sulfuric acid)}$$

Its presence is one reason smog is so irritating to the eyes and to the moist lung passages.

Pure sulfuric acid is a syrupy, colorless liquid, known to the alchemists as "oil of vitriol" (from the Latin *vitrum,* glass; the alchemists thought sulfates looked "glassy"). The properties of sulfuric acid are now embodied in the adjective "vitriolic."

Because sulfuric acid is used in such a variety of industrial processes (textiles, steel, fertilizers, petroleum products), its sales are sometimes used as a barometer of a nation's business. In the United States some 30 million tons are produced each year. Nearly one-third of it goes into the manufacturing of agricultural fertilizers, such as "superphosphate" or ammonium sulfate:

$$H_2SO_4 + 2NH_3 \rightarrow (NH_4)_2SO_4$$

Since it is relatively cheap, sulfuric acid may be used to make other acids:

$$H_2SO_4 + 2NaCl \rightarrow \underset{\text{(hydrochloric acid)}}{2HCl} + Na_2SO_4$$

$$H_2SO_4 + 2NaNO_3 \rightarrow \underset{\text{(nitric acid)}}{2HNO_3} + Na_2SO_4$$

When both hydrogen atoms in the $H_2SO_4$ molecule are replaced by a metal, the compound is called a sulfate. When only one hydrogen is replaced, the compound is called a hydrogen sulfate. The latter still has acidic properties (it can form $H^+$ ions) and is used to clean toilet bowls and drains. Here is how the two salts ionize:

$$\underset{\text{(sodium sulfate)}}{Na_2SO_4} \xrightarrow{HOH} 2Na^+ + SO_4^=$$

$$\underset{\text{(sodium hydrogen sulfate)}}{NaHSO_4} \longrightarrow Na^+ + HSO_4^-$$

$$HSO_4^- \rightarrow H^+ + SO_4^=$$

The presence of the $SO_4^=$ ion in water (from sulfuric acid or dissolved sulfates) can be detected by adding a few drops of a soluble barium (Ba) salt solution. The $Ba^{++}$ forms an insoluble white solid with $SO_4^=$, barium sulfate:

$$Ba^{++} + SO_4^= \rightarrow BaSO_4$$

A final sulfur compound of interest is hydrogen sulfide ($H_2S$), a colorless (and poisonous) gas with an unforgettable rotten-egg odor. It is formed naturally in sewage plants, sulfur springs, the banks of canals and bays, swamps, and in rotten eggs (where sulfur-containing proteins are decomposing).

Despite their obvious differences, sulfur and oxygen are both in group VI A of the Periodic Table and both have outer electron rings with 6 of the 8 possible sites filled with electrons. Each has 2 electron vacancies. For maximum stability, each will require 2 hydrogen atoms, since hydrogen has just 1 electron in its outer shell. This is why there is $H_2O$ but not HO or $H_3O$, and $H_2S$ but not HS or $H_3S$.

The similarity does not end there. Photosynthesis involves $CO_2 + H_2O$, giving carbohydrates and free oxygen. Certain primitive bacteria (*Chlorobium*) photosynthesize in the absence of oxygen by $CO_2 + H_2S$, giving carbohydrates and free sulfur.

Hydrogen sulfide is used in the laboratory to identify unknown metals by forming their sulfides:

$$Zn + H_2S \rightarrow ZnS \text{ (zinc sulfide: white)} + H_2$$
$$Cd + H_2S \rightarrow CdS \text{ (cadmium sulfide: yellow)} + H_2$$
$$2As + 3H_2S \rightarrow As_2S_3 \text{ (arsenic sulfide: lemon yellow)} + 3H_2$$
$$2Sb + 3H_2S \rightarrow Sb_2S_3 \text{ (antimony sulfide: orange)} + 3H_2$$
$$Cu + H_2S \rightarrow CuS \text{ (copper sulfide: black)} + H_2$$
$$Hg + H_2S \rightarrow HgS \text{ (``cinnabar'': red)} + H_2$$
$$Pb + H_2S \rightarrow PbS \text{ (lead sulfide: brown)} + H_2$$

Many of these sulfides are used as pigments (cadmium yellow, for example). It is easy to see why $H_2S$ as an air pollutant discolors white-lead house paint. It also turns silver black (AgS).

*Phosphorus* (*P*). The German alchemist Hennig Brand in 1669 found that when he heated urine with sand and carbon, he obtained a white waxy material that glowed in the dark. He called it phosphorus (lightbearer).

He was really looking for the philosopher's stone, but phosphorus proved almost as lucrative. True to the alchemical code, he refused to tell Robert Boyle how to make it, beyond saying that something from the human body was the source.

Urine contains soluble phosphates; with sand ($SiO_2$), they form phosphorus pentoxide:

| $2Na_3PO_4$ | $+ 3SiO_2$ | $\rightarrow$ | $3Na_2SiO_3$ | $+ P_2O_5$ |
|---|---|---|---|---|
| (sodium phosphate) | (silicon dioxide) | | (sodium silicate) | (phosphorus pentoxide) |

The $P_2O_5$ then reacts with the carbon to form elemental phosphorus:

$$P_2O_5 + 5C \rightarrow 2P + 5CO$$

Phosphorus is in group V A of the Periodic Table, with 5 out of 8 electron sites on the outer shell occupied. Its resemblance to its immediate neighbors, nitrogen and arsenic, is even less than that of sulfur to its neighbors. However, nitrogen and phosphorus are both essential for life, and arsenic is enough like phosphorus to serve as a fatal substitute for it in biochemical reactions.

Calcium phosphate is the main ingredient of bones. Many phosphorus mines (especially in Montana, Utah, Wyoming, and Florida) are actually accumulations of prehistoric bones, dinosaur graveyards.

Calcium phosphate is insoluble and unsatisfactory as a fertilizer. Treatment with sulfuric acid turns it into the more soluble "superphosphate":

$$Ca_3(PO_4)_2 + 2H_2SO_4 \rightarrow Ca(H_2PO_4)_2 + 2CaSO_4$$

(calcium dihydrogen phosphate)

For rapid fertilizing action, ammonium phosphates have the advantage of being very soluble and of ionizing into a double fertilizer ($NH_4^+$ and $PO_4^{\equiv}$). The plant must be able to extract the ions rapidly before they are washed away by rain or irrigation water.

White phosphorus burns easily, so it was used in the first

commercially made matches. Unfortunately, it is highly toxic and the factory workers developed an incurable bone disease. Upon standing (or if heated without air) white phosphorus is changed into red phosphorus, which is more stable. Safety matches contain red phosphorus, not in the match head but in the special striking surface on the box. Friction makes the red phosphorus spark, which ignites the match head.

# 20. The Metalloids

Like those botanical species that merge imperceptibly into one another, the metalloids blend metallic and nonmetallic properties in a way calculated to drive taxonomists mad. Yet to this same area of the Periodic Table belongs carbon, with its rich creative potential.

The three elements that appear below sulfur in group VI A are discussed first:

*Selenium* (*Se*). A grayish solid resembling sulfur in its chemical reactions, but more toxic. Certain soils may contain enough of it to get into plants. Animals grazing on the plants ("locoweed") may develop the "blind staggers" and die.

Before the Plastic Age, bright red automobile taillights were made by adding selenium to the molten glass. Small amounts of selenium are still used to counteract the greenish color of glass, which is caused by iron.

Selenium is used as a semiconductor in transistors. It has the remarkable property of conducting electricity well in light, becoming resistant again in darkness, so it is used in photoelectric cells, burglar alarms, photographers' exposure meters, and automatic door openers.

Selenium sulfide (SeS) is used in antidandruff shampoos.

*Tellurium* (*Te*). A silvery solid, resembling selenium, but more metallic.

The last element in group VIA is *polonium* (*Po*), a radioactive solid discovered and named by the Curies. It is sufficiently metallic to be classified as a metal. Since it ionizes the air around it, it is sometimes attached to phonograph tone-arms as a static charge eliminator.

The remaining metalloids follow:

*Arsenic* (*As*). Located just below phosphorus on the Periodic Table, arsenic enters into biochemical reactions just as phosphorus would but is unable to follow through, so that the process is blocked before completion.

An accurate description of arsenic poisoning was given by the French novelist Gustave Flaubert, the son of a physician, in his masterpiece, *Madame Bovary*. In 1954, Clare Boothe Luce as United States ambassadress to Italy developed mild arsenic poisoning from flakes of paint dropping from her villa ceiling. Arsenic trisulfide ($As_2S_3$) is still used as a pigment in paints.

Arsenic is detected in food (or corpses) by the Marsh test in which the arsenic is converted into the gas arsine ($AsH_3$). When burned, the gas leaves a dark mirror of arsenic on a porcelain dish held in the flame.

The formula of arsine, $AsH_3$, is analogous to that of ammonia, $NH_3$, a reminder that arsenic is in the nitrogen family (group V A).

*Boron (B).* Boron is more like carbon and silicon than it is like the members of its own family (group III A), where its immediate neighbor is aluminum. It is widely distributed over the earth, but in very small amounts. Death Valley, California, provides 85 percent of the world's supply.

Like carbon, boron forms a weak acid ($H_3BO_3$). In fact boric acid is weak enough to be used as an eyewash. The salt, borax ($Na_2B_4O_7 \cdot 10H_2O$), is a useful water softener. It reacts with the $Ca^{++}$ and $Mg^{++}$ ions that make water "hard" and removes them from solution.

When boron oxide ($B_2O_3$) is added to glass, Pyrex glass is obtained. It does not expand when heated as much as ordinary glass does and thus is less likely to crack.

Small amounts of boron are needed for plant growth, but large amounts are toxic, so it is used both in fertilizers and in weed killers.

*Silicon (Si).* If carbon is the key element in the plant and animal world, silicon is the key element in the mineral world. This powdery or crystalline solid is too active to be found free, but combined it makes up roughly one-fourth of the earth's surface.

Sand and quartz are $SiO_2$. Zircon is $ZnO_2 \cdot SiO_2$. Clay is $Al_2O_3 \cdot 2SiO_2 \cdot 2H_2O$. Asbestos, garnet, Portland cement, granite, all are compounds of silicon. When molten minerals are able to cool slowly, a crystalline structure results; when they cool rapidly, crystals do not have time to form and a glassy material results (obsidian, volcanic glass).

Ordinary window glass is made by mixing and heating sand ($SiO_2$), soda ($Na_2CO_3$), and lime ($CaCO_3$). A clear mixture of sodium and calcium silicates results ($Na_2SiO_3 + CaSiO_3$). Other elements can be added for coloring: selenium (ruby), manganese (violet), cobalt (blue).

In natural quartz ($SiO_2$) impurities produce gemstones such as agate, opal, carnelian, amethyst. Carborundum (SiC) is a synthetic abrasive, widely used in emery cloth and grinding wheels.

Made in an electric furnace, carborundum is almost as hard as diamonds.

Silicon is the second member of the carbon family (group IV A) and reacts with other elements in much the same way as carbon does, but the products are quite different—sand ($SiO_2$) and carbon dioxide ($CO_2$), for example. But silicon hydrides (such as $SiH_4$) and hydrocarbons (such as $CH_4$) are analogous. Silicon tetrachloride ($SiCl_4$) and carbon tetrachloride ($CCl_4$) are both low-boiling liquids, but there the similarity ends.

Like carbon (see chapter 21), silicon is able to link up with itself, but the chains are neither as long nor as stable as those of carbon.

Silicon played a minor role in art history. James McNeill Whistler (1834–1903), the painter of the subtle and magical "Nocturnes," began his career as a West Point cadet. He was discharged shortly after he told his chemistry professor that silicon was a gas. In later years Whistler often reflected that "had silicon been a gas, I would have been a major general."

# 21. Carbon: Endless Creation

*. . . for what may be beyond, the eyesight of Lilith is too short. It is enough that there is a beyond.*
——George Bernard Shaw, *Back to Methuselah*

The compounds of all the elements except carbon total about 60,000. The compounds of carbon number between 1,000,000 and 2,000,000. For this reason a special branch of chemistry had to be formed: organic chemistry, which deals exclusively with the compounds of carbon.

The outer electron shell of carbon (first element in group IV A) is exactly half-filled with electrons. It contains 4 electrons and 4 unfilled electron spaces. Thus it can form compounds with metals ($CaC_2$, calcium carbide), or it can act like a metal itself and form compounds with nonmetals (carbon tetrachloride, $CCl_4$). Carbon can form compounds with itself—that is, carbon atoms can join together in a variety of chains and rings to which other elements may attach.

Free carbon is found in more than one form. Only carbon atoms are involved in each case, but they are arranged differently.

Soot or carbon black is carbon formed by incomplete combustion:

$$CH_4 + O_2 \rightarrow 2H_2O + C$$

It is used as a pigment in printer's ink and to make automobile tires black.

Charcoal is similar but is obtained by heating wood in the absence of air. Charcoal is also a useful filtering agent. One cubic centimeter of charcoal (0.4 inch on each side) contains roughly 1,000 square meters of surface (1,200 square yards) for picking up impurities. A charcoal filter in a cigarette will not keep you from getting cancer, but it may slow down the rate at which you get it, since a fraction of all the carcinogenic agents in cigarette smoke will be absorbed by the filter. Charcoal is used to decolorize crude sugar and is generally found in gas masks. It is also a boon to the wine industry. Low-grade natural wines can be run through charcoal until they lose their color and flavor, then treated with flavoring materials and sold under an evocative brand name.

Coal is very old vegetable matter which has broken down into free carbon—that is, it has "carbonized." Fuels of this type can be listed in order of increasing age and carbon content:

| *Fuel* | *Percent carbon* |
|---|---|
| peat | 11 |
| lignite | 22 |
| bituminous coal | 60 |
| anthracite coal | 88 |

When coal is heated in the absence of air, volatile materials are driven off which condense into coal tar, a rich mixture of organic compounds. The carbon and ash that remain are coke, used mainly by the steel industry.

Graphite is a crystalline form of carbon. Its soft flat crystals slide over each other like a deck of cards, making it a good lubricant for dry bearings. It is also the "lead" in lead pencils. Graphite has the metallic property of conducting electricity well; it is used for commutator brushes in electric motors and for battery terminals.

Carbon in the form of coke is converted into graphite by heating it 24 hours at 3,500° C. with iron and silicon oxides. But natural graphite is widely distributed over the world and supplies most industrial needs.

Diamonds are pure crystalline carbon formed by great heat and pressure. In Africa they are found in what seem to be the necks of old volcanos. Although diamonds are the hardest substance known, they will burn if heated in air. If heated to 1,000° C. in the absence of air, they become graphite, not a very profitable transformation. Small diamonds can be made synthetically; they are useful in industry, but not as jewelry.

## Inorganic Carbon Compounds

In addition to the vast numbers of organic compounds, carbon forms simple compounds of great importance. Among these, $CO_2$, $CO$, and $H_2CO_3$ have already been discussed. Two simple salts are derived from carbonic acid ($H_2CO_3$):

$NaHCO_3$ = sodium hydrogen carbonate
$Na_2CO_3$ = sodium carbonate

The first is weakly basic, the second strongly basic. Recalling that bases give $OH^-$ ions to solutions, you may wonder how these two salts can be basic, since there is no OH group in the formulas. While the salts do not add $OH^-$ ions directly, they are able to pull protons ($H^+$ ions) from water molecules, leaving $OH^-$ ions behind:

$$Na_2CO_3 \rightarrow 2Na^+ + CO_3^=$$
$$CO_3^= + HOH \rightarrow HCO_3^- + OH^-$$

$NaHCO_3$ is still known by its old name, sodium bicarbonate. As baking soda or in baking powders it substitutes for yeast as a source of $CO_2$ bubbles to make bread rise.

Carbonate minerals are common, especially those of iron, magnesium, and calcium. Geologists and soil scientists sometimes carry small bottles of hydrochloric acid with them into the field to identify these minerals. A few drops of acid on limestone ($CaCO_3$) gives a vigorous effervescence:

$$CaCO_3 + 2HCl \rightarrow CaCl_2 + H_2O + CO_2\uparrow$$

Since marble is also $CaCO_3$, it becomes clear why acidic smogs are so destructive to marble statues and buildings.

When methane ($CH_4$) from natural gas is treated with chlorine ($Cl_2$), trichloromethane (chloroform) and tetrachloromethane (carbon tetrachloride) are obtained. Both these liquids are rather toxic. Chloroform is seldom used as an anesthetic nowadays. "Carbon tet" is still used as a spot remover, but if too much is inhaled or absorbed through the skin, liver damage will occur. Carbon tetrachloride is also used as a fire extinguisher when electrical equipment is involved. Being a nonconductor, it will not cause damaging short circuits, as water might.

The solid ionic salt, calcium carbide ($CaC_2$), hydrolyzes in water to form acetylene:

$$CaC_2 + 2H_2O \rightarrow Ca(OH)_2 + C_2H_2\uparrow$$

This is the basis of the Coleman lanterns used by campers. Acetylene is a starting material for many organic syntheses. Mixed with oxygen it gives the oxy-acetylene flame for welding and cutting metals.

With nitrogen, carbon forms the cyanides, highly toxic, but still used in gold mining, photography, and electroplating. With almost any acid, cyanides release hydrocyanic acid (HCN), a colorless gas with the famous "odor of bitter almonds":

$$KCN + KCl \rightarrow KCl + HCN$$

An extremely fast-acting poison, HCN stops all tissue respiration and paralyzes the central nervous system. A dose of 0.05 gram (0.002 ounce) is fatal. It has played a part in countless real and fictional crimes. It was the lethal agent of the California gas chamber.

## Radiocarbon Dating

As has been explained, atomic weights are relative; they compare the weight of a given element with the weight of the element used as a standard (originally hydrogen). So atomic weights should be simple whole numbers. The weights are due to the numbers of protons and neutrons in the atoms nucleus, and these particles weigh the same no matter what element they are in. There should be no fragments or fractions of particles in atomic nuclei.

Yet when one consults a table of atomic weights, one finds that the atomic weight of chlorine is 35.453, carbon is 12.011, and so forth. This is because the atomic weights are *averages*. They are averages of all the naturally occurring isotopes of the given elements. As described in chapter 17, the isotopes of an element all have the same chemical properties, but they have different atomic weights due to different numbers of neutrons in their nuclei. For example, here are the three isotopes of carbon:

| *Isotope* | *Percent in nature* | *Number of protons* | *Number of neutrons* | *Total mass* |
|---|---|---|---|---|
| carbon-12 | 98.89 | 6 | 6 | 12 |
| carbon-13 | 1.11 | 6 | 7 | 13 |
| carbon-14 | trace | 6 | 8 | 14 |

Thus carbon with a weight of 12 predominates in nature, but there are enough of the higher-weight isotopes to push the average atomic weight up to 12.011.

Carbon-14 is continually produced in the upper atmosphere by the action of cosmic rays. Radioactive, it has a half-life of 5600 years (after 5600 years, 50 percent has disintegrated).

Carbon-containing materials, including living organisms, are in equilibrium with the three carbon isotopes and show the same percentages of each as in the preceding table. Their radioactivity from the carbon-14 they contain is 16 beta-disintegrations per minute per gram of carbon.

When a carbon-containing material is buried or sealed off from the atmosphere, it is no longer in equilibrium with carbon-14. The carbon-14 it contains slowly disappears and the beta-disintegrations decline. When the material is 5600 years old, half of its carbon-14 will be gone and its beta-disintegrations will have dropped to 8.

Thus it is possible to relate the beta-radiation of a carbonaceous material to the age of that material. If an archaeologist digs up an Indian sandal, he can have its beta-disintegration rate determined and learn the age of the sandal to within 200 years. Egyptian mummies, papyri, Stonehenge, the Dead Sea scrolls, coal and oil deposits have all been "dated" in this way.

## Organic Chemistry

After chemistry became firmly established as a science, organic chemistry and biochemistry developed as two successive waves. Their gloriously intricate and fragile structures could only be built on an unshakable base of accurate chemical information. Even though the great creative flowering of organic chemistry, between 1860 and 1910, is long past, research activity continues to be most active in organic chemistry and biochemistry.

All this activity is possible because of a unique property of the carbon atom. Most of the time, instead of losing or gaining electrons, it *shares* electrons. Here is what happens when carbon reacts with hydrogen to form methane:

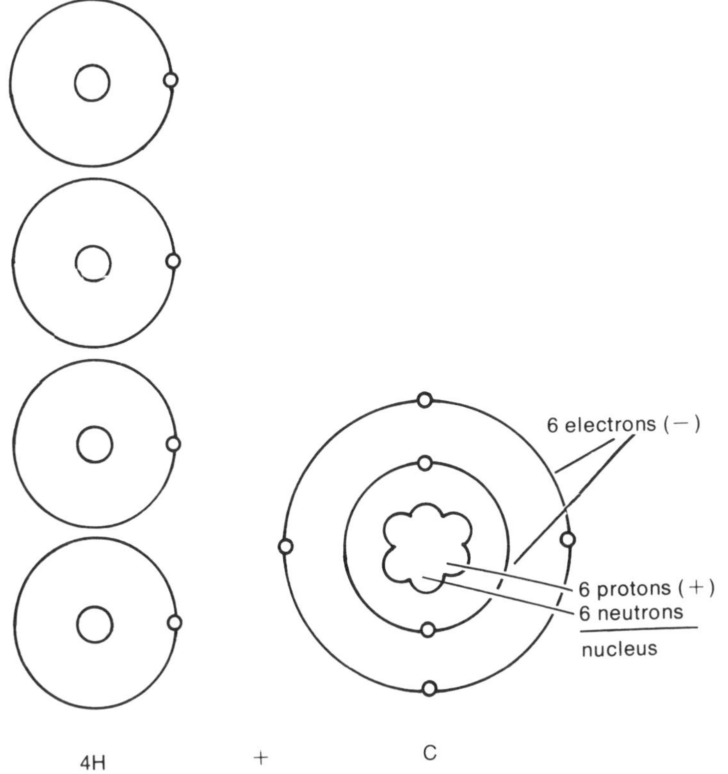
6 electrons (−)
6 protons (+)
6 neutrons
nucleus
4H
+
C

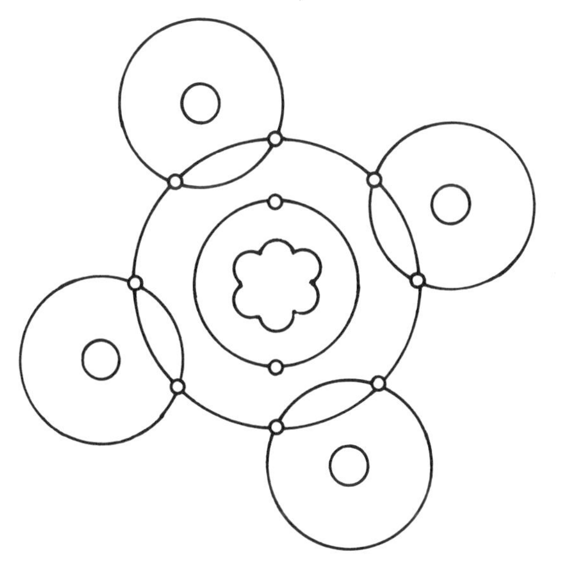

By sharing electrons, each element is able to complete its outer electron ring (2 for hydrogen, 8 for carbon). A simpler way of representing this is by using chemical symbols with a line indicating a pair of shared electrons. Methane then becomes:

$$\begin{array}{ccccc} & & H & & \\ & & | & & \\ H & - & C & - & H \\ & & | & & \\ & & H & & \end{array}$$

And when carbon shares electrons with itself, compounds like the gas propane:

$$\begin{array}{ccccc} & & H & & \\ & & | & & \\ H_3C & - & C & - & CH_3 \\ & & | & & \\ & & H & & \end{array}$$

and the liquid octane:

$$\begin{array}{ccccccccccccccc} & & H & & H & & H & & H & & H & & H & & \\ & & | & & | & & | & & | & & | & & | & & \\ H_3C & - & C & - & C & - & C & - & C & - & C & - & C & - & CH_3 \\ & & | & & | & & | & & | & & | & & | & & \\ & & H & & H & & H & & H & & H & & H & & \end{array}$$

are formed. When the carbon atoms are linked in chains like this, they are called aliphatic ("oily") compounds. When carbon atoms are linked together in rings, they are called aromatic ("smelly") compounds. Benzene is a common example:

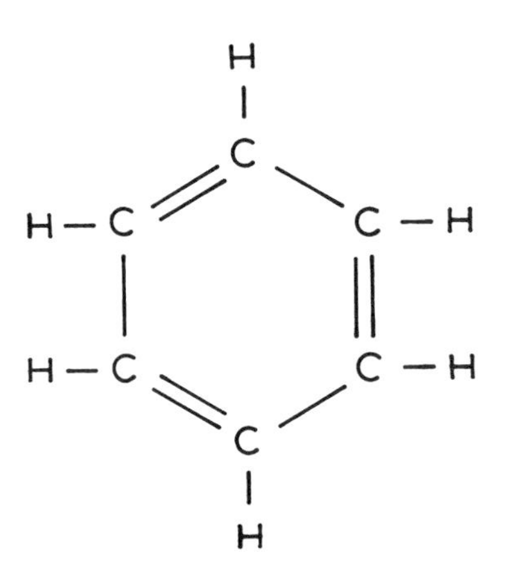

which is often simplified to

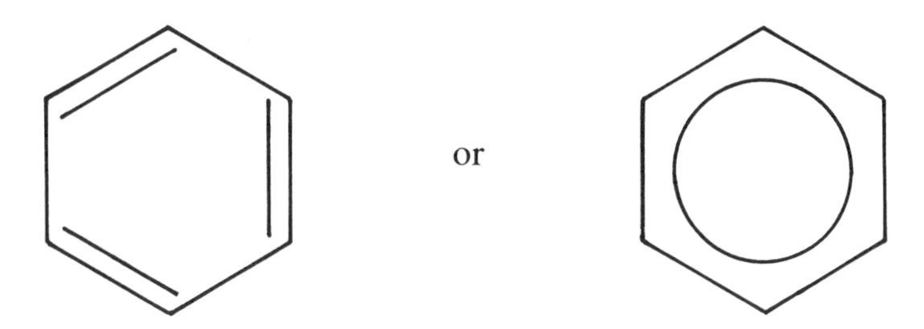


Remember that each of the six angles represents a C atom with an H atom attached. If the figure is drawn as:

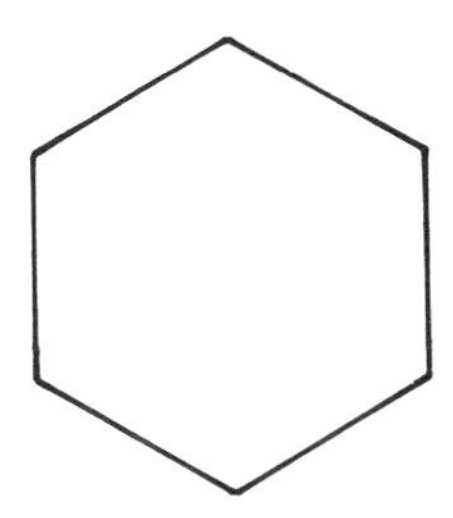

then two H atoms are attached to the C atom and the compound is cyclohexane:

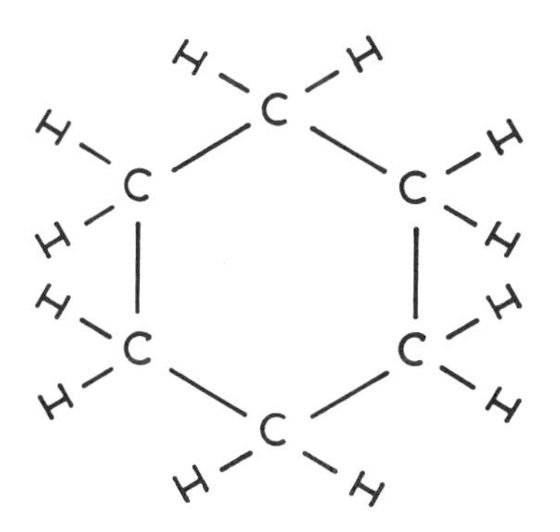


Simplified ring formulas were devised by chemists so that instead of representing each H atom they could represent complex formulas like the one on the next page.

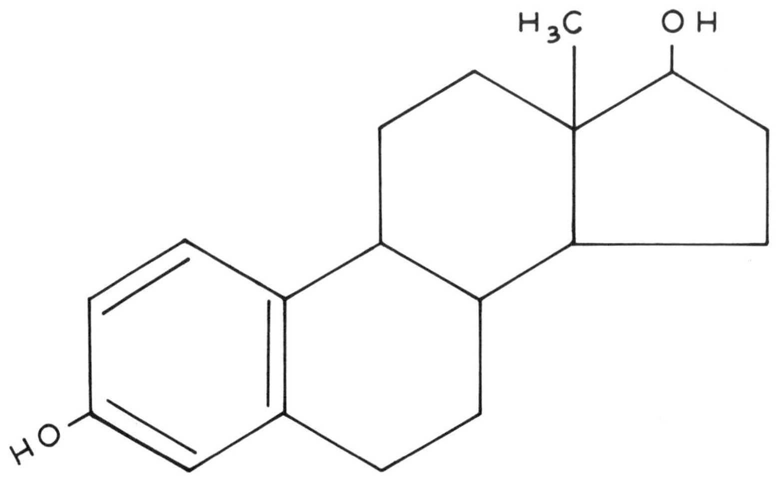


estradiol, a female sex hormone

A different element may substitute for one or more of the C atoms in a ring compound; there may be more or less than six atoms in the ring:

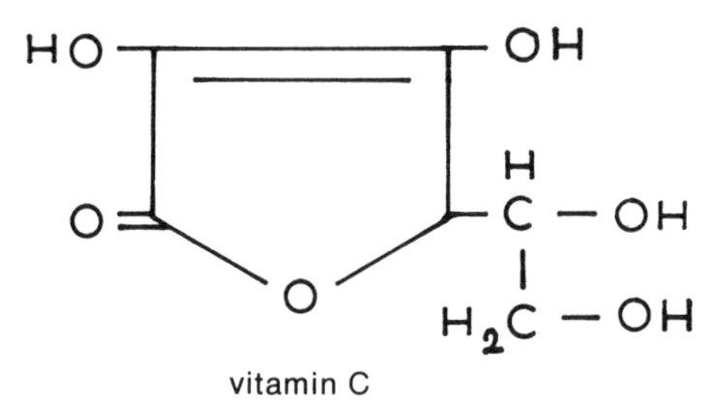


vitamin C

The ring may have side chains (see illustration on following page).

These are only two-dimensional schemata, not the three-dimensional reality, but they may give a hint of the variety possible in organic compounds.

Chemists have synthesized many compounds not found in nature (barbiturates, plastics, pesticides), but in a surprising number of cases they have merely copied or made slight improvements in compounds that already exist in nature (aspirin, LSD, penicillin). The skeleton of all these compounds is carbon,

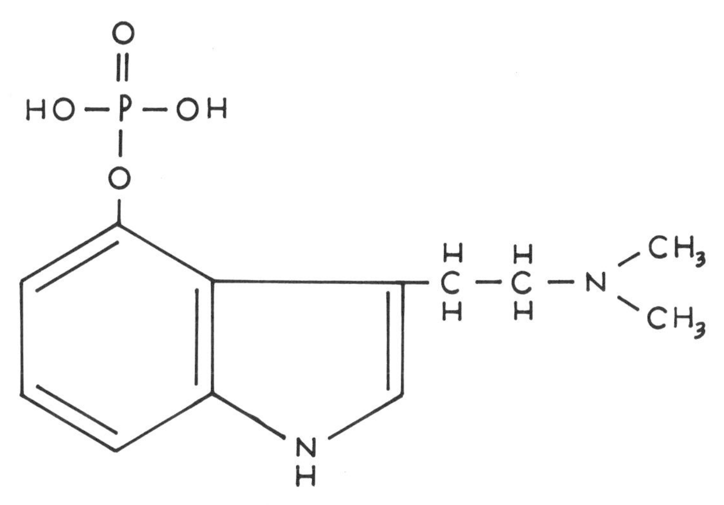


psilocybin, a mushroom hallucinogen

just as the skeleton of all the compounds in living organisms is carbon.

Plants and animals are watery bags of carbon compounds. Carbohydrates, fats, proteins, hormones, and genetic material are all carbon compounds. Even the enzymes by which they utilize their carbon compound nutrients are themselves carbon compounds.

There is a Hindu tradition that nature is more playful than purposeful and that the universe is simply the "play" (*lila*) of God. If so, the carbon atom must be one of the things He plays with most. No other element could give physical expression to an infinitely creative imagination.

# Glossary

# Suggestions for Further Reading

# Index

# Glossary

ACID A PROTON donor; a substance which produces hydrogen IONS ($H^+$) in solution.

ALCHEMY The art or science, practiced since before the birth of Christ, which seeks to transmute baser metals into gold (exoteric alchemy) and to achieve personal enlightenment and immortality (esoteric alchemy).

ALPHA PARTICLE The nucleus of a helium atom, 2 neutrons and 2 protons, positively charged.

ALPHA RAYS Streams of fast-moving helium nuclei.

ATOM The smallest particle of an ELEMENT that retains all of its chemical properties.

ATOMIC NUCLEUS The dense core of an atom consisting of neutrons (uncharged) and protons (positively charged).

ATOMIC NUMBER The number of protons or the number of electrons associated with a given atom.

ATOMIC WEIGHT The relative weights of the atoms of an ELEMENT compared with those of carbon-12, which has been given the value 12.

AVOGADRO'S PRINCIPLE Equal volumes of all gases at the same temperature and pressure contain the same number of molecules.

BASE A PROTON acceptor; a substance which reacts with an ACID to form a SALT.

BETA PARTICLE An ELECTRON emitted during the radioactive decay of an atomic nucleus.

BETA RAYS Streams of electrons from radioactive nuclei, traveling nearly as fast as light and resembling CATHODE rays.

BOILING Rapid evaporation from within and from the surface of a liquid. Occurs at a specific temperature called the boiling point, which is characteristic for each liquid.

CATHODE The negative electrode of a circuit, battery, or vacuum tube from which ELECTRONS issue.

CELSIUS (C). TEMPERATURE SCALE The centigrade temperature scale in which the freezing point of water is 0° C. and its boiling point is 100° C.

CENTI- A prefix meaning 1/100, used in the METRIC SYSTEM. A centimeter is 1/100 of a meter.

CHEMICAL CHANGE A change in the composition, and often in the energy content, of a substance.

CHEMICAL COMPOUND A combination of two or more ELEMENTS in a definite proportion by weight: pure, homogeneous, and having properties distinct from the elements that compose it.

CHEMICAL PROPERTIES Characteristics of a substance concerned with its composition, energy content, and reactivity.

CHEMICAL REACTION A process in which the ATOMS of one MOLECULE increase, decrease, or rearrange themselves in a reciprocal fashion with those of another chemical agent.

CHEMISTRY The science that investigates the composition and the properties of the substances composing the universe.

CONDENSATION A change in state from vapor to liquid, the opposite of EVAPORATION.

CONDUCTOR Any material through which ELECTRONS flow easily when an electrical force is applied.

DENSITY The MASS of a substance per unit VOLUME.

DEUTERIUM A hydrogen ISOTOPE occurring naturally in water and having a mass number of 2.

ELECTRIC CURRENT The rate of ELECTRON flow, in which energy is carried from one place to another.

ELECTRON A part of an atom that orbits the nucleus, has a mass of 1/1837 of a hydrogen atom, and is negatively charged.

ELEMENT A substance composed of ATOMS all of which have the same ATOMIC NUMBER and therefore the same CHEMICAL PROPERTIES.

ENERGY The ability or capacity to do work.

EVAPORATION A change in state from liquid to vapor taking place at the surface of the liquid (which is cooled thereby).

FAHRENHEIT (F.) TEMPERATURE SCALE A temperature scale in which the freezing point of water is 32° F. and its boiling point is 212° F.

FREEZING A change in state from liquid to solid form, releasing energy; the opposite of MELTING.

GAMMA RAYS High-frequency electromagnetic radiation emitted by the nuclei of radioactive atoms, similar to X RAYS.

GROUP One of the vertical columns of elements in the PERIODIC TABLE.

HALF-LIFE The time required for half the atoms of a radioactive element to decay.

HEAT The total molecular energies in a body, usually measured in calories.

HYDROLYSIS The splitting of a compound by a reaction with water.

INORGANIC SUBSTANCE Any of the elements and their compounds except the compounds of carbon.

ION An atom or group of atoms which has lost one or more electrons (a cation, positively charged), or gained one or more electrons (an anion, negatively charged).

ISOTOPES Atoms of the same element whose nuclei have the same number of protons but different numbers of neutrons and therefore different nuclear masses.

KILO- A prefix meaning 1,000, used in the METRIC SYSTEM. A kilogram is 1,000 grams.

KINETIC ENERGY The energy from motion.

LITER The volume of 1 kilogram of water at 4° C., used in the METRIC SYSTEM.

MASS The amount of matter in a body, not affected by changes in gravitational pull, but measured by the extent to which it resists efforts made to start it or stop it.

MATTER That which has mass and occupies space.

MELTING A change in state from solid to liquid, with absorption of energy.

METRIC SYSTEM A system of measurement developed in France during the 1790s, one of the enduring achievements of the French Revolution. Ten and multiples of ten are used exclusively. The meter, defined as 1/10,000,000 of the distance from the equator to the North Pole, was its original standard. Since then the meter has been redefined in terms of the wavelength of light emitted by krypton-86. Thus the standard is available to scientists all over the world, rather than immured on a platinum bar at 0° C. at the Bureau of Standards in Paris.

Universally adopted by scientists, the metric system is also used commercially in most countries. The United States is the most notable exception; it still supports the awkward "British system." The major approximate equivalents are:

| | | | |
|---|---|---|---|
| Length: | 1 centimeter | = | 0.4 inch |
| | 1 meter | = | 1.1 yard |
| | 1 kilometer | = | 0.6 mile |
| Volume: | 1 liter | = | 1.06 quart |
| Weight: | 1 kilogram | = | 2.2 pounds |

MICRO- A prefix meaning 1/1,000,000, used in the METRIC SYSTEM. A microgram is 1/1,000,000 of a gram.

MILLI- A prefix meaning 1/1,000, used in the METRIC SYSTEM. A milligram is 1/1,000 of a gram.

MOLAR (M) The strength of a solution expressed in terms of moles per liter of the dissolved substance. A bottle labeled "5M NaOH" contains a 5-molar solution of sodium hydroxide.

MOLE The weight in grams of the atomic or molecular weight of a substance. The molecular weight of NaOH is 40. A mole of NaOH is 40 grams. A 1-molar solution of NaOH is 40 grams of NaOH in a liter of water.

MOLECULAR FORMULA A formula stating the kinds and numbers of atoms present in a molecule. The molecular formula of sulfuric acid is $H_2SO_4$.

MOLECULAR WEIGHT The sum of the weights of all the atoms in a molecule.

MOLECULE The smallest particle of a compound that has all its chemical properties; two or more atoms (identical or different) held together in chemical combination.

NITROGEN FIXATION The process of converting atmospheric nitrogen into a compound that plants and other living organisms can use.

NUCLEAR FISSION The splitting of the nucleus of a heavy atom into two main parts, accompanied by the release of much energy. This was the basis of the atomic bombs dropped on Hiroshima and Nagasaki in 1945.

NUCLEAR FUSION The combination of the nuclei of light atoms to form heavier nuclei, accompanied by the release of much energy. This is the basis of the hydrogen (thermonuclear) bomb, first tested in 1952.

ORGANIC SUBSTANCES Compounds of carbon, whether produced by living organisms or synthetically.

PERIOD One of the horizontal rows of elements in the PERIODIC TABLE.

PERIODIC LAW When the ELEMENTS are arranged in the order of their increasing ATOMIC NUMBERS, a periodic recurrence of properties can be observed.

PERIODIC TABLE An arrangement of the chemical elements in order of their increasing atomic numbers so as to show a periodic recurrence of properties.

PHYSICAL CHANGE A change in the physical state of a substance with no change in its composition. The melting of ice is a physical change.

PHYSICAL PROPERTIES Characteristics of a substance not involving changes in its composition—solubility, melting point, and so forth.

PRESSURE Force per unit area, expressed as number of pounds per square inch, grams per square centimeter, or in similar terms.

PROTON The nucleus of a hydrogen atom; a hydrogen ION; a positively charged subatomic particle.

QUANTUM THEORY A general physical theory that energy is radiated in discrete units called quanta and that material particles have wavelike properties.

RADIOACTIVE ISOTOPE An unstable ISOTOPE of an element which decomposes and gives off alpha, beta, gamma, or other radiation.

RADIOACTIVITY Term suggested by Marie Curie for the disintegration of atomic nuclei in which alpha, beta, or gamma radiation is produced.

SALT A compound made up of a metallic (+) ION from a BASE and a nonmetallic (−) ION from an ACID; for example, NaCl from NaOH ($Na^+$) and HCl ($Cl^-$).

SOLUTE A dissolved substance.

SOLUTION A homogenous mixture of the MOLECULES, ATOMS, or IONS of two or more substances.

SOLVENT A dissolving substance. Water is the most common solvent.

SPECTROSCOPE An optical instrument that separates light or radiation from any source into its constituent frequencies. Spectra so produced can be analyzed to determine the condition and the composition of the bodies from which they originated.

SPONTANEOUS COMBUSTION An exothermic reaction so rapid that the kindling temperature of one of the reactants is reached.

STEEL Iron alloyed with smaller amounts of other metals and carbon to obtain greater strength or other properties.

"S.T.P." Standard Temperature and Pressure, or standard conditions. Since the volume of a gas is affected by temperature and pressure, these variables must be fixed when volume is being determined. A pressure of 1 atmosphere (the average atmospheric pressure at sea level) and a temperature of 0° C. are the standard conditions employed.

STRUCTURAL FORMULA A formula showing which atoms in a molecule are bound to which other atoms and by what kinds of bond. The structural formula of sulfuric acid is:

```
         O
         ↑
HO ——— S ——— OH
         ↓
         O
```

SUBLIMATION The direct conversion of a substance from the solid state to the vapor state (or sometimes the reverse process), without passing through the liquid state.

TEMPERATURE A measure of the average KINETIC ENERGY per MOLECULE in a body, expressed in degrees (Celsius, Fahrenheit, or Kelvin scale).

TITRATION A process of determining the strength of an unknown solution (usually an acid or a base) by adding a solution of known strength which reacts with it.

TRANSMUTATION Conversion of the ATOMIC NUCLEUS of one ELE-

MENT into an atomic nucleus of another element through a loss or gain in the number of PROTONS.

TRITIUM The heaviest ISOTOPE of hydrogen, mass number 3, produced by nuclear reactors.

UNSATURATED The condition of an organic compound when it has double (C=C) or triple (C≡C) bonds between some of its atoms, so that it reacts readily with hydrogen or other substances.

VALENCE A number that shows the combining power of an ELEMENT, usually the number of ELECTRONS the element loses, gains, or shares.

VOLUME The quantity of space a body occupies.

WEIGHT The force which gravity exerts on a material body.

X RAYS Electromagnetic rays of very short wavelength, produced when a beam of ELECTRONS strikes a material object.

# Suggestions for Further Reading

Books on chemistry are currently available in great variety, many of them in paperback. Those listed here are a limited selection: books of sustained usefulness, representative books, or books with particular interest or charm.

HISTORICAL BOOKS

Crosland, Maurice P. *Historical Studies in the Language of Chemistry*. Cambridge, Mass.: Harvard University Press, 1962. The horrendous problems of chemical terminology from the alchemists through Lavoisier and Dalton are analyzed in a fresh and engaging fashion.

Holmyard, E. J. *Alchemy*. Baltimore: Penguin Books, 1968 (Pelican; first published 1957). Still the best and most complete of the many books on alchemy. The author's attitude toward his material is ideally balanced, neither gullible nor patronizing.

Ihde, Aaron J. *The Development of Modern Chemistry*. New York: Harper & Row, 1964. A comprehensive treatment, especially thorough for the late nineteenth and the twentieth centuries. The bibliography, presented as a running commentary, is more useful and informative than conventional listings.

Partington, J. R. *A History of Chemistry*. 4 vols. New York: St. Martin, 1962–1971. Not only the most exhaustively detailed and complete work in the field, but one of tremendous interest.

———. *A Short History of Chemistry*. New York: Harper & Row, 1960 (Harper Torchbook T 522).

TEXTBOOKS

Dickerson, Richard E.; Gray, Harry B.; and Haight, Gilbert P., Jr. *Chemical Principles*. New York: W. A. Benjamin, 1970. A college text for science majors, remarkable for the skill with which it makes complex topics understandable.

Pimental, George C. (ed.). *Chemistry—An Experimental Science*. San Francisco: W. H. Freeman, 1963. This text, written collectively by twenty high school teachers, shows what a high standard such books can reach.

Williams, Arthur L.; Embree, Harland D.; and DeBey, Harold J. *Introduction to Chemistry*. Reading, Mass.: Addison-Wesley, 1968. For students not majoring in chemistry, this covers inorganic chemistry, organic chemistry, and biochemistry. Highly readable and well suited to self-instruction.

POPULAR BOOKS

Farber, Eduard (ed.). *Great Chemists*. New York: Interscience Publishers, 1970. Includes lives of more than 100 chemists. The individual pieces vary in comprehensiveness but the 1600 pages contain a great deal of information.

Jaffe, Bernard. *Crucibles: The Story of Chemistry*. New York: Simon & Schuster, 1930, rev. 1957 (paperback, rev. and abridged. New York: Fawcett, 1960). This book is to chemistry what Paul de Kruif's *Microbe Hunters* is to bacteriology: undocumented, melodramatic, supremely readable. A score of key figures in chemistry have been brought to life more vividly here than anywhere else, and occasional lapses in scholarship or taste cannot obscure that achievement.

Lapp, Ralph E., and the Editors of *Life*. *Matter*. New York: Time, Inc., 1963. This beautifully illustrated book features a periodic table with color photographs of all the elements. The text, though highly compressed, contains at least one sentence on every major topic.

LIBRARY GUIDE

Bottle, R. T. (ed.). *The Use of Chemical Literature*. 2nd ed. Hamden, Conn.: Archon, Shoe String Press, 1969. The vast store of chemical information in libraries sometimes baffles even chemists. This guide supplies the background, purpose, and limitations of reference works, as well as suggestions and practical advice from experts.

# Index

# About the Author

Richard Furnald Smith received his B.S. degree in Chemistry in 1950, his M.S. in Biochemistry in 1952, and his Ph.D. in Biochemistry in 1962, all from the University of Arizona. He was formerly a research biochemist with the Forest Service of the United States Department of Agriculture and is currently teaching at the University of California Extension, Berkeley, California. *Chemistry for the Million* is his first book and is based on lectures presented at Laney College in Oakland, California.